# Japan
Second edition

**The World's Landscapes**
Edited by Dr J. M. Houston

Australia
Brazil
China
France
Ireland (Second edition)
**Japan (Second edition)**
New Zealand
South Africa
The Soviet Union (Second edition)
Terrain Evaluation
Wales

# Japan: geographical background to urban-industrial development

Second edition

David Kornhauser
Professor of Geography, University of Hawaii

with a Foreword by

J. M. Houston
Chancellor of Regent College, Vancouver

**Longman**
London and New York

**Longman Group Limited**
Longman House
Burnt Mill, Harlow, Essex CM20 2JE, England
*Associated companies throughout the world.*

*Published in the United States of America
by Longman Inc., New York*

© Longman Group Limited 1976, 1982

All rights reserved; no part of this publication may be reproduced, stored in a retrieval system, or transmitted in any form or by any means, electronic, mechanical, photocopying, recording, or otherwise, without the prior written permission of the Publishers.

*First published in 1976
Second impression 1977
Second edition 1982
Second impression 1985*

---

**British Library Cataloguing in Publication Data**

Kornhauser, David
 Japan: geographical background to
 urban-industrial development. —2nd ed.
 —(The World's landscapes)
 1. Cities and towns—Japan
 I. Title II. Kornhauser, David. Urban Japan
 III. Series
 307.7'6'0952    GF125
 ISBN 0-582-30081-9

**Library of Congress Cataloging in Publication Data**

Kornhauser, David Henry, 1918—
 Japan, geographical background to urban-industrial development.

 (The World's landscapes)
 Rev. ed. of: Urban Japan. 1976.
 Bibliography: p.
 Includes index.
 1. Anthropo-geography—Japan. 2. Cities and towns—
Japan. 3. Japan—Description and travel. 4. Japan—
Economic conditions. I. Title. II. Series.
GF125.K67 1982    952    81—19394
ISBN 0—582—30081—9    AACR2

---

Produced by Longman Singapore Publishers Pte Ltd.
Printed in Singapore

# Contents

| | |
|---|---|
| List of photographs | vii |
| List of tables | ix |
| List of maps | x |
| Foreword by Dr J. M. Houston | xi |
| Introduction | xiii |

**1. The urban landscape** — 1
  I. Physical components — 1
    Landforms — 4
    Organisation of surface features — 5
    Coasts — 8
    Lowlands, drainage and elements of weather and climate — 10
    Comparative size — 15
    Natural resources and city development — 17
  II. Some human components of the urban landscape — 22
    Population and settlement — 22
    Population density — 23
    Ways of life — 26
    Changes in types of settlement — 30
    Some political aspects of contemporary urbanisation — 32

**2. The agrarian landscape** — 34
  Background to the present — 34
  Elements of agricultural evolution — 37
  Edo period developments — 40
  Late Tokugawa period developments — 42
  Agriculture after 1868 — 42
  Agricultural growth since 1945 — 46
  Postwar land reform — 48
  Some contemporary problems of agriculture — 53
  Summary — 55

**3. The city in Japanese history** — 58
  The beginnings: AD 710 to 1572 — 58
  The establishment of a permanent network of cities: 1573 to 1868 — 64
    Ports — 72
    Post or stage towns — 72
    Religious centres — 74
    Market towns — 74
    Spas — 74
  Summary — 75

*Contents*

| | |
|---|---|
| **4. Changes in the urban landscape after 1868** | 77 |
| Some effects of transportation on city growth and change | 78 |
| Transportation after 1868 | 79 |
| Further effects of transportation on cities | 82 |
| The functional transformation of historic cities | 84 |
|     Castle towns | 84 |
|     Port cities | 87 |
|     Stage towns | 87 |
|     Religious centres | 88 |
|     Resorts | 88 |
| Summary | 91 |
| **5. Historical aspects of the commercial landscape** | 94 |
| Commerce and trade in the Edo period | 97 |
| Edo period industry | 101 |
| Summary | 104 |
| **6. Landscapes of commerce and industry, 1868–1945** | 107 |
| Industrial developments, 1850–67 | 108 |
| The modernisation of the industrial landscape | 112 |
| Some effects of the war on the industrial landscape | 121 |
| Summary | 122 |
| **7. Reconstruction and the growth of the contemporary industrial landscape** | 125 |
| The dual industrial structure | 131 |
| Major modern industries and landscape change | 134 |
| A new industrial landscape | 137 |
| Summary | 141 |
| **8. The Japanese landscape** | 144 |
| Natural setting and human adaptation | 144 |
| The agrarian landscape and agriculture | 145 |
| The urban landscape | 148 |
| The industrial and commercial landscape | 150 |
| Urban, metropolitan and regional planning | 153 |
| Planning and future prospects | 158 |
| **Notes** | 163 |
| **Bibliography** | 168 |
| **Index** | 174 |

# List of photographs

| | | |
|---|---|---|
| 1.1 | Coastal landscape, western Shikoku | 5 |
| 1.2 | Crater of Mount Aso in northcentral Kyushu | 9 |
| 1.3 | Oyster fisheries in Hiroshima Bay | 20 |
| 1.4 | Congested settlement at Sakashitsu | 25 |
| 2.1 | Factories in fields | 35 |
| 2.2a | Fertilised seedlings being prepared for germination | 39 |
| b | Seedlings being transplanted by machine | 39 |
| 2.3a | Computer rooms in hydroponic farm near Nagoya | 44 |
| b | Vegetables being grown in computer-controlled hydroponic greenhouse | 44 |
| 2.4 | Modern farmhouse, Iga Basin | 46 |
| 2.5 | Rice harvesting combine | 51 |
| 2.6 | Family harvesting by combine | 55 |
| 3.1 | Canalised distributary of the Yodo River at Osaka | 62 |
| 3.2 | Osaka Castle | 65 |
| 3.3 | Kiso Fukushima, an old post town | 73 |
| 4.1 | Toll highway, Chiyoda-ku, Tokyo | 82 |
| 4.2 | Overview of the same highway | 82 |
| 4.3 | High-rise buildings in the Shinjuku district of Tokyo | 83 |
| 4.4 | Apartment and single-family dwellings in Sendai | 87 |
| 4.5 | Large-scale *danchi*, Tama New Town, Tokyo | 90 |
| 5.1 | Used-car dealer, Sendai by-pass | 100 |
| 5.2 | Drive-in restaurant, Kanagawa Ken | 101 |
| 5.3 | People boarding minibus near Tama New Town, Tokyo | 103 |
| 5.4 | Noodle stand, Kanagawa prefecture | 104 |
| 6.1 | Small family housing, Sendai | 109 |
| 6.2a | Housing in upper-middle-class residential suburb, Nishi | |
| b | Kamakura, near Yokohama | 110 |
| 6.3 | Mitsubishi shipyards, Nagasaki | 116 |
| 6.4 | Main street leading from railway station, Sendai | 121 |
| 7.1 | Coal- and petroleum-powered thermal electric plant, Shiogama | 126 |
| 7.2 | Industry and urban sprawl in the northern margins of Hiroshima | 128 |

*List of photographs*

| | | |
|---|---|---|
| 7.3 | Land newly reclaimed for industrial use along the foreshore of the Seto Inland sea | 139 |
| 7.4 | Sunday on a main street in the northern part of central Tokyo | 141 |
| 8.1 | New prefectural hospital near the border of Shizuoka and Kanagawa Ken | 146 |
| 8.2 | Neighbourhood department store, Tama New Town, Tokyo | 149 |
| 8.3 | Car ferry, Sendai | 152 |
| 8.4 | The Ginza, Tokyo | 159 |
| 8.5 | Mall in front of Sendai's new railway station | 160 |
| 8.6 | Crowd on sidewalk opposite Sendai station | 161 |

# List of tables

| | | |
|---|---|---|
| Table 1.1 | Areas and percentages of urban and non-urban land in Japan, 1920–75, and index of change | 3 |
| Table 1.2 | Total area in slope of less than 15° | 13 |
| Table 1.3 | General population growth 1875–1970 = quinquennially; 1971–79 = yearly | 24 |
| Table 1.4 | National population by region, 1980 | 26 |
| Table 1.5a | Densely inhabited district populations encircling the largest cities, 1975 | 27 |
| Table 1.5b | Areas of densely inhabited districts encircling the largest cities, 1975 | 27 |
| Table 1.6 | Employed persons by types of industry, 1960–75 | 28 |
| Table 1.7 | Percentages of population in cities and counties, 1920–75 | 29 |
| Table 1.8 | Changes in numbers and types of settlement, 1883–1978 | 30 |
| Table 2.1 | Percentages of population in cities, towns and villages by size group, 1940–50 | 48 |
| Table 3.1 | Castles, built or repaired, 1575–1616, and present city status | 68 |

# List of maps

| | | |
|---|---|---|
| Map 1 | Japan: place names and physical features | xvi |
| Map 2 | Geomorphology of Japan: tectonic divisions | 7 |
| Map 3 | The tectonic divisions simplified to show the structural arcs and nodes in the Japanese Archipelago | 8 |
| Map 4 | Coastal types of Japan | 10 |
| Map 5 | Terrain character of Japan | 11 |
| Map 6 | Atmospheric circulation, January and July | 13 |
| Map 7 | Landforms of East Asia, showing distribution of lowland | 14 |
| Map 8 | Japan: industrial belt and urban core | 16 |
| Map 9 | Road, rail and communications of Japan | 18 |
| Map 10 | Ocean currents in the vicinity of Japan | 19 |
| Map 11 | Districts, subdistricts and prefectures in Japan | 31 |
| Map 12 | Major cities and travel routes in Japan during the Edo period | 63 |
| Map 13 | Industrial city zones | 155 |
| Map 14 | Six new special areas for industrialisation | 156 |

# Foreword
by Dr J. M. Houston, Chancellor of Regent College, Vancouver

Despite the multitude of geographical books that deal with differing areas of the world, no series has before attempted to explain man's role in moulding and changing its diverse landscapes. At the most there are books that study individual areas in detail, but usually in language too technical for the general reader. It is the purpose of this series to take regional geographical studies to the frontiers of contemporary research on the making of the world's landscapes. This is being done by specialists, each in his own area, yet in non-technical language that should appeal to both the general reader and to the discerning student.

We are leaving behind us an age that has viewed nature as an objective reality. Today, we are living in a more pragmatic, less idealistic age. The nouns of previous thought forms are the verbs of a new outlook. Pure thought is being replaced by the use of knowledge for a technological society, busily engaged in changing the face of the earth. It is an age of operational thinking. The very functions of nature are being threatened by scientific takeovers, and it is not too fanciful to predict that the daily weather, the biological cycles of life processes, as well as the energy of the atom will become harnessed to human corporations. Thus it becomes imperative that all thoughtful citizens of our world today should know something of the changes man has already wrought in his physical habitat, and which he is now modifying with accelerating power.

Studies of man's impact on the landscapes of the earth are expanding rapidly. They involve diverse disciplines such as Quaternary sciences, archaeology, history and anthropology, with subjects that range from pollen analysis to plant domestication, field systems, settlement patterns and industrial land use. But with his sense of place, and his sympathy for synthesis, the geographer is well placed to handle this diversity of data in a meaningful manner. The appraisal of landscape changes, how and when man has altered and remoulded the surface of the earth, is both pragmatic and interesting to a wide range of readers.

The concept of 'landscape' is of course both concrete and elusive. In its Anglo-Saxon origin, *landscipe* referred to some unit of area that was a natural entity, such as the lands of a tribe or of a feudal lord. It was only at the end of the sixteenth century that, through the influence of Dutch landscape painters, the word also acquired the idea of a unit of visual

## Foreword

perceptions, of a view. In the German *landschaft*, both definitions have been maintained, a source of confusion and uncertainty in the use of the term. However, despite scholarly analysis of its ambiguity, the concept of landscape has increasing currency precisely because of its ambiguity. It refers to the total man—land complex in place and time, suggesting spatial interactions, and indicative of visual features that we can select, such as field and settlement patterns, set in the mosaics of relief, soils and vegetation. Thus the 'landscape' is the point of reference in the selection of widely ranging data. It is the tangible context of man's association with the earth. It is the documentary evidence of the power of human perception to mould the resources of nature into human usage, a perception as varied as his cultures. Today, the ideological attitudes of man are being more dramatically imprinted on the earth than ever before, owing to his technological capabilities.

In this original and penetrating study of Japan, the emphasis is on the urban-industrial scene. This is a corrective to the popular impressions of outsiders who think of Japan as a landscape of gardens in cherry blossom. It behoves the treatment of the third largest industrial nation of the world to take its urban life seriously, when some 110 million people are concentrated in a restricted area. However, the evolution of the agrarian landscapes is given careful study and the importance of the seventeenth century is given due emphasis; man has paid a heavy price for the apparent docility of the Japanese landscapes. It is also important to note the cultural influence of China on Japan, not only for the cult of the garden but for the initial rise of the city. Although Japan has long roots in the past, the book has a significant interest in the contemporary changes in the urban and rural landscapes. As in the past, there is every indication that in future foreign elements will not dominate, but will exert a selective influence: the emerging landscapes of Japan will retain their distinctive Japanese character.

J. M. HOUSTON

Since the publication of the first edition, the role of Japan as a leading urban-industrial nation has been enhanced and despite the energy crisis of the early 1970s, which caused other industrial economies to falter, the Japanese drive towards a technologically advanced and economically prosperous society has continued. Yet there have been costs, as well as benefits, and these are reflected in the modern Japanese landscape. Professor Kornhauser's book, carefully and fully revised and brought up-to-date, remains an essential introduction for the student and thoughtful layperson who seeks a well-informed and attractive explanation of the complexities of the modern Japan as reflected in its landscapes.

# Introduction

Discussion of the geographic features of any single part of the world may be viewed as an attempt to employ with some legitimacy the term *region*, and the past failure to do this with any scientific exactness has caused a generation of geographers to argue against the regional approach. On the other hand, man through time has often adapted himself to specific segments of the earth in relatively singular fashion, to form distinctive enclaves of culture, and in many ways to shape his physical environment with similar distinctiveness. Perhaps in such gross terms as landforms, geomorphology, climate, soils, drainage, vegetation. and natural resources, one part of the earth bears too close a relationship with others to warrant separate consideration, at least in any scientifically measurable sense; but in such attributes as human history, ways of looking at, of doing and shaping things, the region may indeed be unlike any other. Furthermore, the tendency for one cultural group to inhabit a particular part of the earth and to evolve a milieu that can be immediately distinguished as separate and unique is nowhere more clearly represented than in the physical and cultural landscape of Japan.

I have chosen, in discussing the Japanese landscape, to emphasise the urban-industrial scene; this should be no cause for surprise, considering that Japan today is one of the three leading industrial nations of the world, and that its people are in the main classified as urban. Japan is the first non-western culture to have reached a kind of post-industrial stage of development, yet because the Japanese themselves so often stress the importance of tradition in the life of the people — although much of this, I feel, is only perfunctory — there may still be some confusion among non-Japanese about the extent of the transformation of the culture from basic agrarian (as recently as 1945), to the present state of urban-industrial sophistication. The present volume is one attempt to remedy any misunderstanding on this point.

The list of those to whom I am indebted for help in this effort is far too long for adequate mention here. My special thanks are to my old friend and companion-in-the-field, Professor Norton S. Ginsburg, who originally proposed the venture; and to Professor Philip N. Jenner of the University of Hawaii. Dr Jenner and I have been close associates and friends since army language-school days during the Second World War,

*Introduction*

and realising his language skills and his unerring literary tastes, not to mention his understanding of Japan and of me, I have been most fortunate to have had him act as an informal editor. He has devoted himself meticulously and unstintingly despite a heavy schedule of his own, and any merit this work may have is heavily attributable to his efforts.

Among others who have had a hand in this, either directly or indirectly, are my Japanese colleague, Professor Tatsuo Ito of Mie University, who read and commented on the first chapter; Dr Shinzo Kiuchi, the preeminent Japanese urban geographer, who has offered invaluable advice through the years; Professor Toshio Noh, my old friend of Michigan and Sendai days; the late Curtis A. Manchester, Japan specialist and former colleague at Hawaii, and Professor Richard J. Pearson of the University of British Columbia. Douglas Gordon, who has extensive knowledge of Japan, while a member of this department read the entire manuscript and gave much valuable criticism, as did the late Professor Richard K. Beardsley, my friend and former teacher, both at Michigan and in Japan.

I wish to thank also the Ministry of Education of Japan, the Japan Center for Area Development Research, the Japan Society for the Promotion of Science, and the Office of Research Administration of the University of Hawaii, for providing funds for long periods of research in and travel to Japan. The Bureau of Statistics of the Office of the Prime Minister of the Japanese Government, and the Tokyo Institute for Municipal Research have contributed precious research materials over the years. I am also grateful for advice and encouragement to Professor Dr Peter Schöller of the Ruhr University at Bochum, to my former colleagues at Tokyo Kyoiku (now Tsukuba National) University, to those at Tohoku and Mie Universities in Japan and lastly to my friends and academic associates, Dr William J. Chambliss of the University of Kentucky and Professor Ardath W. Burks, the well-known political scientist and Japan specialist of Rutgers, the State University of New Jersey.

I wish to dedicate this second edition to my beloved wife, Michiko, who, in all matters pertaining to Japan and to me, has helped without reserve.

> DAVID H. KORNHAUSER
> *University of Hawaii*
> *Honolulu*
>
> *Spring 1981*

Map 1  Japan: place names and physical features

# 1
# The urban landscape

## I. Physical components

Japan seems relatively easy to define as a distinctive cultural entity. It is insular in the extreme, being at least 160 kilometres from the Korean peninsula, the closest projection of the Eurasian landmass. Its culture, particularly its language, although heavily influenced by the Chinese, especially in the early stages of its recorded history, is often so singular as to defy classifications which attempt to group cultural elements on a global basis.[1] Moreover, the special qualities of Japanese culture were vastly enhanced and ramified by roughly 250 years of political isolation — a period that ended only in the 1860s, so recently that when a major course of modernisation was finally launched in the 1870s the Industrial Revolution had already caused basic cultural changes in much of the mid-latitude world. The same kinds of changes that have occurred in other cultures since then, such as the rapid reduction of the primary in favour of the secondary and tertiary industrial sectors, have become pervasive in Japan only since the mid-1950s.

The landscape of Japan, along with the philosophical and religious influences, some introduced, especially from China in the first massive waves of cultural importation after the sixth century AD, has traditionally been an important force in producing in the Japanese people an extraordinary aesthetic and artistic sensitivity for which they have long been admired, especially in the West. The original beauty of their natural surroundings was evidently such as to inspire early Japanese not only to express their appreciation in art and literature, but even, in their daily tasks, carefully to nurture and protect the charm and grandeur of their natural environment, despite the slow transformation of the landscape from the purely natural to one largely shaped by man. Mostly of course, such change has been associated with an agrarian way of life.

At present, with the development of a truly widespread urban-industrial landscape, for the first time in their history the former dominance of the scene by agrarian ways is being seriously threatened. Such a challenge is not new in the Japanese experience. Accommodation to new influences, often accepted indiscriminately and later modified, blended and rendered distinctly Japanese, is, as will be mentioned hereafter, a recurrent theme in

## The urban landscape

the nearly 1 500 years of Japanese recorded history. The moulding of a new and different landscape is now occurring rapidly in Japan, and often with consequences that seem out of keeping with tradition. In time, the remarkable talent of the Japanese for accommodation and selective change, which has so often seen foreign elements softened into a pleasing and thoroughly Japanese picture, should come to the fore in the creation of a landscape whose features may be dissimilar and alien to those of the past.

Transition of the landscape from the dominance of the traditional-rural, to that of the modern urban-industrial, is extremely new in Japan. Much of the obvious change in the countryside has occurred since 1950 (and probably more recently), while the pace of change underwent great acceleration in the 1960s, as a result of unusually rapid economic growth and advancing technology. In an attempt to present Japan realistically as the highly developed urban-industrial culture it has become, it seems logical to accent the modern landscape, with the traditional and basic physical elements as a background.

Only a few decades ago, and certainly before the Second World War, the Japanese landscape, even by statistical division into 'urban' and 'rural' (Table 1.1), was overwhelmingly rural. Conversion to other categories had been relatively insignificant. Such change as occurred involved mainly the slow creation of new level areas, usually to enlarge the agricultural domain predominantly for the production of wet rice, either by terracing slope lands or by reclaiming shallow off-shore features. In addition, certain areas — flat lands for the most part — were converted for urban, or at least for non-agricultural use; but as of 1940, the aggregate space so committed was less than 3 per cent of the total (Table 1.1). On the other hand, since nearly three-quarters of the land surface exceeds 15° in average slope (Table 1.2), even this small amount represented a sizeable portion that has grown larger with each census since 1920. From less than 0.5 per cent in 1920, land in cities by 1950 had surpassed 5 per cent, and in 1975, the most recent general source for this information, the figure had reached 27 per cent of the total surface, albeit there was a sudden increase following the 1953 law to amalgamate towns and villages into cities, about which more will be said hereafter.

Conversion of the landscape for agriculture and forestry is, of course, another matter to be approached, mainly in Chapter 2. This process has been taking place in Japan since before the start of written history in the sixth century, and has involved the bulk of level or gently sloping land for wet rice as the principal crop, and most of the hill and mountain surfaces for dry field crops and trees, at least in 'Old Japan', a general term for all of Japan except Hokkaido and Okinawa. The conversion of Hokkaido landscapes did not begin until after the Meiji Restoration of 1868, and is therefore well behind the rest of the nation in intensity, despite rich production of wet rice and other crops in many low-lying areas since that time. In the main, the transformation of the Japanese landscape, particularly of 'Old Japan', has been so complete that the expression 'man-made'

TABLE 1.1  Areas and percentages of urban and non-urban land in Japan, 1920–75, and index of change

| Year | Areas (km²) and percentages | | | | | | | Index of change since 1920 | |
|---|---|---|---|---|---|---|---|---|---|
| | Total* | Cities | % | Counties | % | % in dispute | | Cities | Counties |
| 1975 | 377 534.99 | 102 409.77 | 27.13 | 273 963.17 | 72.57 | 0.30 | | 7 446.03 | 72.01 |
| 1970 | 377 308.68 | 95 382.76 | 25.28 | 280 693.52 | 74.39 | 0.33 | | 6 935.11 | 73.78 |
| 1965 | 377 267.17 | 88 573.33 | 23.48 | 287 269.04 | 76.15 | 0.37 | | 6 440.01 | 75.51 |
| 1960 | 377 151.08 | 82 903.74 | 21.98 | 292 801.00 | 77.64 | 0.38 | | 6 027.78 | 76.97 |
| 1955 | 377 151.08 | 67 979.67 | 18.03 | 307 870.85 | 81.63 | 0.34 | | 4 942.68 | 80.93 |
| 1950 | 377 099.08 | 20 031.26 | 5.31 | 356 925.64 | 94.65 | 0.04 | | 1 456.44 | 93.82 |
| 1947 | 377 284.79 | 16 110.40 | 4.27 | 361 174.39 | 95.73 | — | | 1 171.36 | 94.94 |
| 1945 | 377 297.97 | 14 571.19 | 3.86 | 362 726.78 | 96.14 | — | | 1 059.45 | 95.35 |
| 1940 | 382 545.42 | 8 852.01 | 2.31 | 373 693.41 | 97.69 | — | | 643.61 | 98.23 |
| 1935 | 382 545.42 | 5 094.53 | 1.33 | 377 450.89 | 98.67 | — | | 370.41 | 99.22 |
| 1930 | 382 264.91 | 2 950.65 | 0.77 | 379 314.26 | 99.23 | — | | 214.54 | 99.71 |
| 1925 | 381 810.06 | 2 181.50 | 0.57 | 379 628.57 | 99.43 | — | | 158.61 | 99.79 |
| 1920 | 381 808.04 | 1 375.36 | 0.36 | 380 432.68 | 99.64 | — | | 100.00 | 100.00 |

Source: 1975 Population Census of Japan, vol. 1, Total Population, Table 3, p. 2.
*The sudden decrease in total land after 1940 is attributable to the loss of northern islands to the USSR after 1945, and the increase in city area after 1950 was caused by the 1953 legislation ordering the amalgamation of administrative areas.

*The urban landscape*

is sometimes used when referring to the traditional scene.[2]

Since it is clear, not only in such figures as are presented in Table 1.1, but also in statistics pertaining to changing human endeavour (Table 1.6), that the Japanese landscape is moving rapidly in the direction of even visual dominance by the city and its effects, this presentation at the outset will emphasise the urban scene, with the general and relatively constant physical elements as a background.

**Landforms**

The landforms of Japan would not seem to lend themselves readily to the growth of a predominantly urban culture, yet, as will be explored in Chapter 3, city life has deep historical roots and owes its beginnings to the very early settlement in key lowlands. These, although few and often far between, are situated close to prominent sea lanes and frequently one to another. The Japanese have tended to organise lowland associations and eventually to consolidate this development into the thriving milieu that has in many respects attained megalopolitan proportions. Much of this has come about because of the extraordinary ability of the Japanese in human and especially in political organisation, but the physical background has been a powerful factor, and an understanding of its rudiments is essential to any discussion of the present.

The concentration of humans and their works in a series of narrow lowlands, and the gradual emergence, between modern Tokyo in the east and Osaka in the west, of a *core* area of increasingly intensive land use, has been in response not only to physical configuration and climate, but equally to history and the modern demands of commerce and trade. The gross structure of the region is such that the broadest and hence most useful plains have generally been built contiguous to the shores of the Pacific and its chief Japanese appendage, the Setonaikai (Seto Inland Sea) which, happily, are heavily indented, allowing ample access to sea lanes.[3] It is noteworthy that the most important international shipping routes of the present day have consistently tended to lie in the same position.

The largely volcanic backbone that projects from the sea floor, whose exposed tips form the Japanese Archipelago, tends to be arranged in arcuate patterns, with prominent nodes in northern Kyushu, central Honshu, and central Hokkaido, each of which represents a point of contact with external arcs such as the Kuril chain in the northeast, the Ryukyu Archipelago in the south and the Bonin Island group in the southeast. As previously mentioned, of the four main islands of Japan, Kyushu in the south, which is nearly linked to the largest island, Honshu in the centre, and Shikoku, smallest of the group, are culturally considered as 'Old Japan', while Hokkaido, the northern island and second in size, if no longer a genuine frontier, is at least a kind of outland because of the comparative newness of its exploitation. The total area of roughly 370 000 square kilometres is more than 10 per cent smaller than that of the state of

*Physical components*

1.1 Coastal landscape, western Shikoku: Sakashitsu, a small fishing port, facing the Bungo Straits between Shikoku and Kyushu, west of the city of Uwajima

California, with which Japan is so often compared in size. The islands of Sado in the Sea of Japan near the city of Niigata; Awaji, at the eastern side of the Setonaikai close to Kobe, the Amakusa island group, west of Kumamoto Prefecture in western Kyushu, and Okinawa in the far south, are next in size within the islands of Japan. All have shared in the general cultural development and are readily identified on the map.

These islands and the hundreds of smaller ones that lie near them are in themselves often arc-shaped, and the whole archipelago is a giant arc, similar to many Pacific coastal features of eastern Asia in that between the arc and the continent is a relatively shallow sea (the Sea of Japan in this case), and off the convex side of the arc is a great depression in the earth's crust (the Japan Trench). This condition is not only seen in Asia, or exclusively in land and sea associations, it is especially noteworthy across the Eurasian landmass and around the entire Pacific rim, and is present in many parts of the world, characteristically wherever orogeny has been comparatively recent.

Conditions of arcuateness, involving a fairly shallow feature (often a sea) and a deep crustal trench with protrusions lying between (in this situation, the islands of Japan), protrusions that are still in a generally active tectonic state, usually indicate instability and such manifestations as earthquakes and hot springs, for all of which Japan is famous.

**Organisation of surface features**

The important mountain systems of Japan are often located in close association with a series of major fault lines. The most prominent of these lies across southwestern Japan from the Kii Peninsula of southern Honshu through Shikoku to northcentral Kyushu, in the west. Less obvious is the line of dislocation, from central Honshu northward, and opinions differ as to the general structure of the area.[4] Specialists generally agree that the southwestern part of Japan is tectonically divided by the so-called 'Median Dislocation Line', into an inner and an outer portion, with the former having become more important in human development by virtue of its gentler slopes and more numerous (usually deltaic) lowlands, as well as by a more equable climate, both conditions being most marked around the Setonaikai. In the northeast, some accept this kind of broad physiographic regionalisation, while others disagree, but it is generally recognised that the Pacific flank of northern Honshu is composed of older rocks, and the Sea of Japan flank of younger and more volcanic materials.[5]

The mountain systems tend to follow the basic arcuate pattern, with a fundamental structural interruption in central Honshu by a gigantic and now much ramified tectonic cone known as the *Fossa Magna*, which crosses Honshu more or less at right angles to the southwest to northeast mountain trends, from the vicinity of the Izu Peninsula in the south, to east of the Noto Peninsula on the Sea of Japan in the north. This area is spectacularly punctuated by volcanic features created subsequently to the initial block-faulting which created the great rift (the *Fossa Magna*), and includes the incomparable zone of Fuji, whose height of 3 776 metres (12 388 ft) is supreme in the realm. In terms of surface form today, however, it is now thought that the major terrain boundary of this area is not the *Fossa Magna*, but the great central knot of Honshu, the Chubu Node.[6] In any case, the geomorphic patterns of the whole region and their relationships to present configuration are so complex that the reader is referred to more complete descriptions elsewhere.

The Setonaikai is an inundated structural depression, flanked by the high mountains of Shikoku in the south (which parallel the Median Dislocation Line), and by the somewhat less rugged Chugoku Highlands of western Honshu in the north.

In the northeast, the linear structure of the mountain systems continues, with three components in Honshu, the Ou and Dewa Ranges, of north—south alignment, in centre and west, respectively; and in the east, the largely non-volcanic Kitakami and Abukuma Highlands, a single upland feature, separated by the Sendai Plain. These highland features, especially in the volcanic portions, are interspersed with or flanked by shallow, intermontane basins or by mainly north—south aligned river valleys, as well as by several prominent coastal lowlands along the Sea of Japan.

The structure of Hokkaido is dominated by two features, the great Hokkaido Node in the centre of the island, representing the area of contact between the mountains of this area and the external arcs of Sakhalin to the north and the Kuril chain to the east; and by the broad, productive lowlands of the Ishikari and other rivers west of the node. This

## Physical components

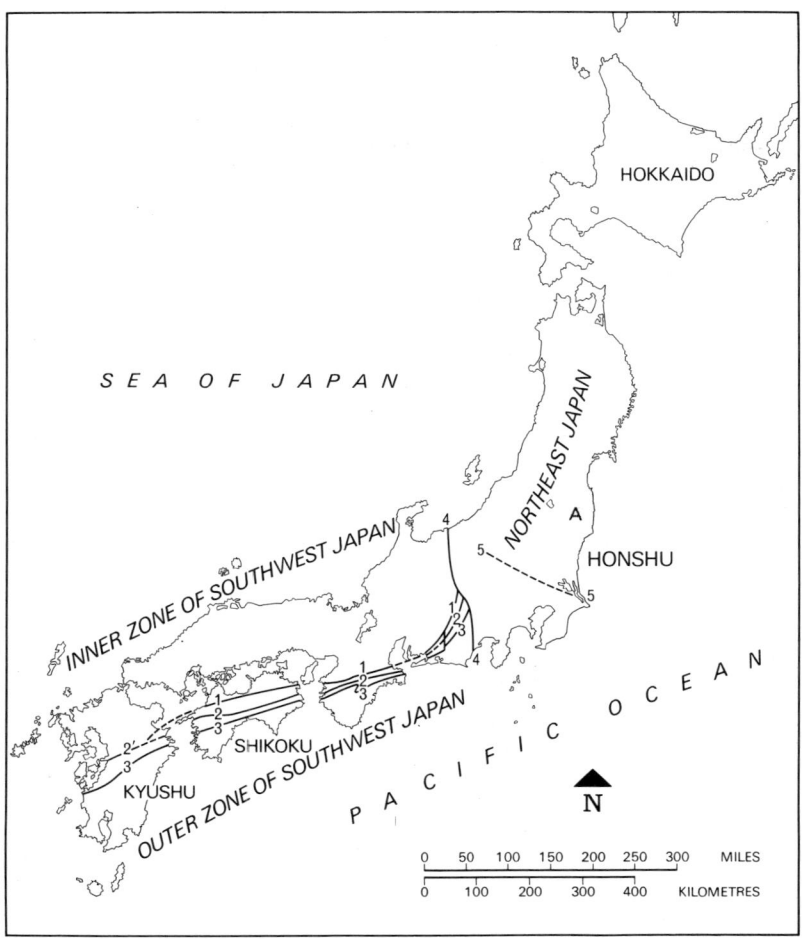

Map 2 Geomorphology of Japan: tectonic divisions. (*Source*: Takai *et al*, 1963)

1: Median dislocation (tectonic) line
2: Mikabu line
2': Usuki-Yatsushiro line
3: Butsuzo line

4: Itoigawa-Shizuoka line (Fossa Magna)
5: Kanto tectonic line

A: Abukuma metamorphics

latter lowland tends to divide the region into a rough rhomboidal portion, marked by the Hokkaido Node and its appendages, and into a southwestern 'tail', whose mountains are more akin to the western ranges of northern Honshu.

Kyushu is equally complex, but its chief lineaments are the northcentral

## The urban landscape

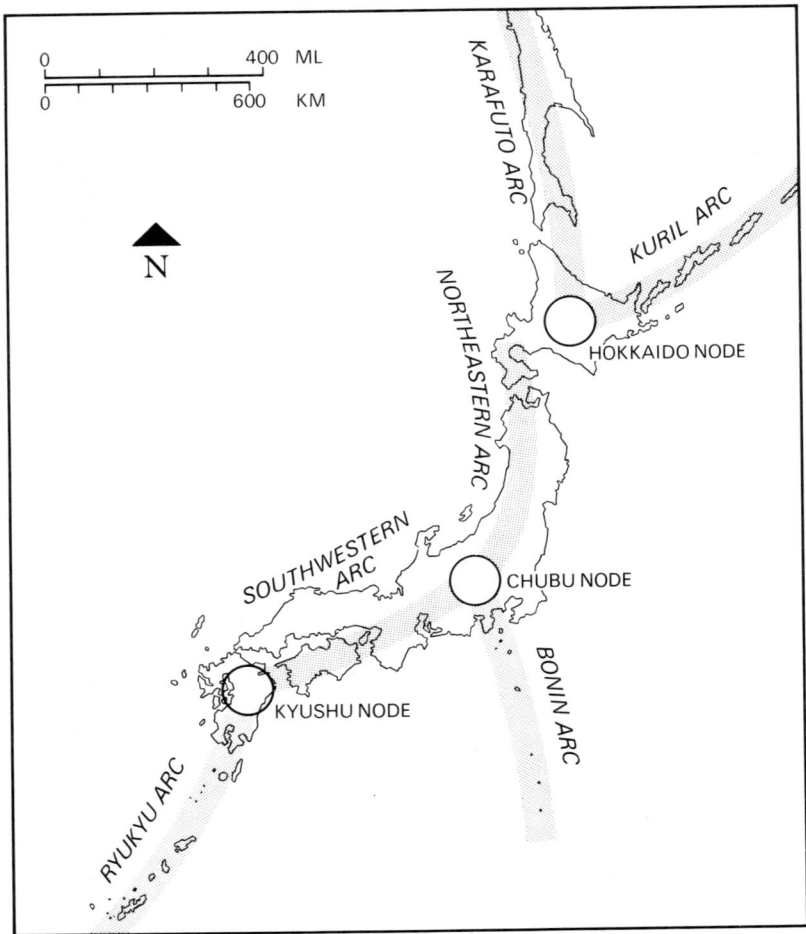

Map 3  The tectonic divisions (see Map 2) simplified to show the structural arcs and nodes of the Japanese Archipelago. Six main arcs are involved, and three principal nodal areas. (*Source*: Trewartha, 1965)

Kyushu Node, associated with Mount Aso, the world's largest *caldera*, and the point of coalescence between the external Ryukyu arc and the southwestern mountain arc of Japan proper. Kyushu is therefore, as mentioned above, divided into an inner zone of broader lowlands, gentler climate and denser settlement; and an outer zone of more rugged topography, more climatic uncertainty, and less human development.

### Coasts

As noted, the coastline of Japan is a major natural asset. Trewartha asserts

*Physical components*

1.2 Crater of Mount Aso in northcentral Kyushu, the world's largest *caldera*-type volcano

that the ratio between sea frontage and coast averages 1 linear kilometre for each 8.5 square kilometres of land in Japan, compared, for example, to about 1:13 square kilometres for Great Britain. He also suggests in this regard, that Japan can be divided into three parts: a northeastern region where uplift has created coastal terraces, coastal plains, and dissected fans; an intermediate region where elevation and depression have about equally influenced coastal forms, and a southwestern region, mainly around the Setonaikai and in northwestern Kyushu, where subsidence has dominated.[7]

The result has been that the Pacific side has become the more variegated, with the major bays being, from east to west, Tokyo, Sagami, Suruga, Ise, and Osaka, and including also the great Setonaikai and its chief openings to the Pacific, the Bungo Channel in the west, and the Kii Channel in the east.

Except in northwestern Kyushu, the Sea of Japan littoral has only a few important indentations. These occur particularly around Wakasa Bay, in westcentral Honshu and at Toyama Bay, east of the Noto peninsula. On the Pacific side northeast of Tokyo coastal indentations are many, but since these are usually backed by rugged highlands, especially north of Sendai along the famous *ria* coast, the westernmost Pacific coastal embayments are the most useful to man.

In general, however, the aforementioned major bays along the Pacific are not the best natural harbours. The lowlands associated with them and forming the core area of human development have therefore had to rely on man-made port facilities or on other artificial means to acquire appropriate access to the Pacific. Even the deepest embayments on the Sea of Japan

*The urban landscape*

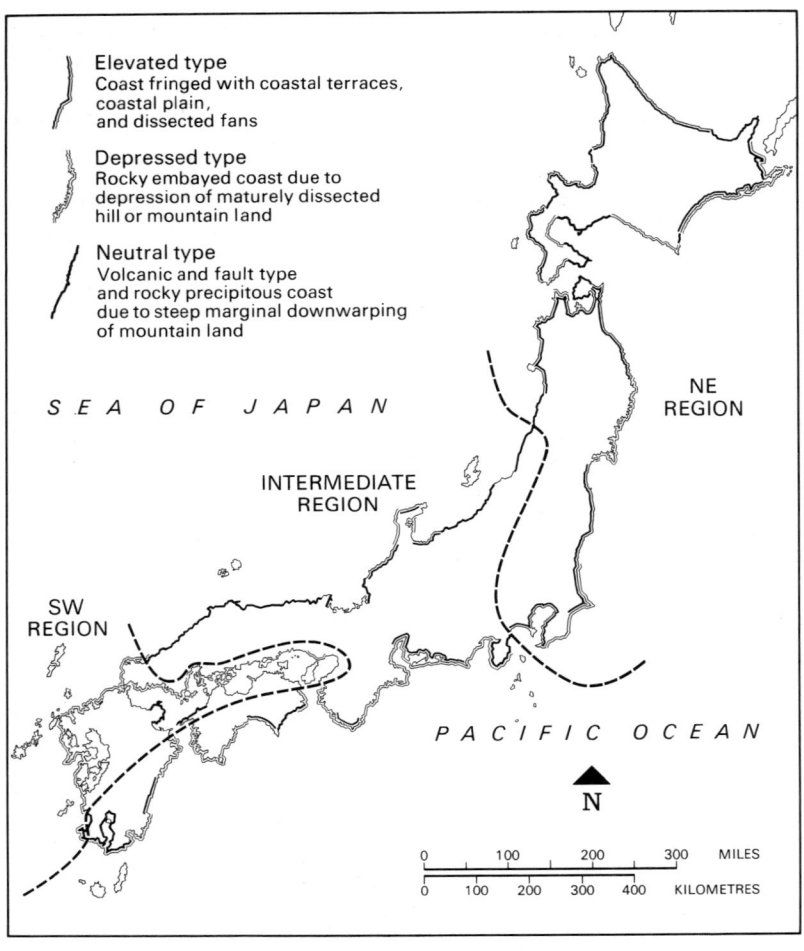

Map 4 Coastal types of Japan. (*Source*: Trewartha, 1965)

coast are not the best harbours, so that elaborate refashioning of river mouths has been necessary to provide maritime access for such places as the important city of Niigata.

### Lowlands, drainage and elements of weather and climate

In terms of their contribution to the development of culture, especially the culture of the present, the major plains of Japan are of such importance as to demand separate treatment. They are important also in a much broader sense, as a rather extreme example of man's ability to utilise

*Physical components*

almost to the fullest the most meagre benefits of his environment, and indeed to make these into a centre of one of the largest and most vital population blocks of the contemporary world.

The major plains of Japan, as elsewhere, are the products of initial and continuous earth movement, coupled with such atmospheric forces as weathering, erosion, and the climatic cycle, to which may be added, with special emphasis in this case, the reshaping of lowland surfaces by man. As previously noted, landscapes have been altered through the centuries, first by having been systematically converted into irrigable and hence *level* plots for the cultivation primarily of wet rice, and more recently by man's

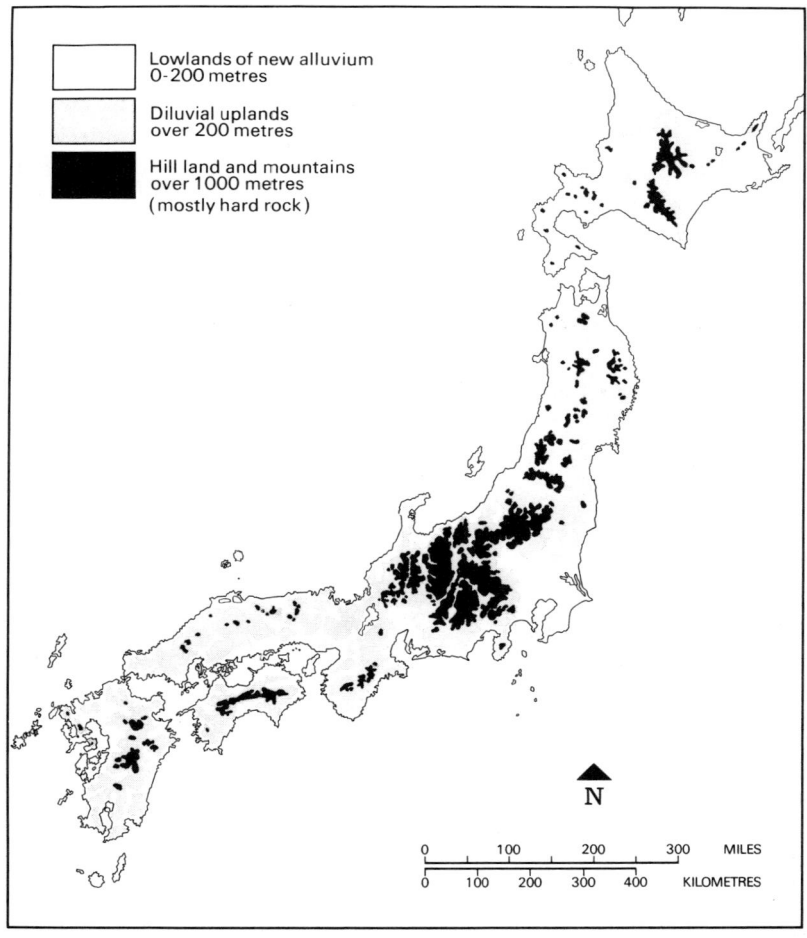

Map 5 Terrain character of Japan. (*Source*: Trewartha, 1965)

rapidly growing technical ability to transform the environment for such needs as housing and industry.

The plains themselves were formed in their present locations by the directions of prevailing winds and weather, acting on and slowly reducing the arcuate mountain systems that form the backbone of the archipelago. Very broadly, the atmospheric cycle is dominated by an east Asian monsoon of winter, which, though initially dry and extremely cool, is warmed and humidified as it crosses the open water of the Sea of Japan; and by a Pacific maritime air mass, or oceanic monsoon, of summer. Consequently Japan generally experiences a climatic pattern similar to that of the east coast of North America, though more modified by marine influence. For most of the year, these two air masses, supplemented by gentle, cyclonic storms of continental origin in late winter and early spring, and by heavy rains accompanying typhoons in late summer and early fall, are associated with surface winds that prevail from the northwest in winter and from the southeast in summer. As a result of these winds and of the generally year-round precipitation, which is much affected by the orientation of the orographic (mountain) barriers, a pattern of drainage has formed which features major stream flow toward the Pacific coasts of central Honshu, and toward the shores of the Setonaikai. Thus the mountain systems, which tend to descend sharply to other coasts, have been graded in these areas into frequently more gentle contours, albeit the total area under $15°$ of slope is remarkably small (Table 1.2). The Japanese have simply organised this meagre area by constructing an increasingly interlocking series of concentrated settlements; as activity has advanced those living outside have been constrained to shift their operations, and more recently their residences, towards the centre. There are exceptions, but mostly these relate to peripheral zones such as the area around Sapporo in Hokkaido or the Sendai Plain in the northeast; or to places where local conditions are narrowly favourable, as in specific lowlands along the coasts of the Sea of Japan, or in inland basins.

It would be convenient to state with certainty the percentage of level land in Japan and to show population densities in these areas, but unfortunately the situation is too complex. For one thing, the lowlands are discontinuous, and the presence of man and his effects is felt not only here but also in adjacent highlands. For another, the sizes of the plains are subject to controversy and change, and probably cannot be determined precisely. Table 1.2 gives a rough indication of these circumstances, and later tables concerned with population, which relate urban prefectures to population features, may also be considered to indicate very general conditions of land use intensity in the major plains.

All Japanese plains are exceptionally small when compared with other of the world's great lowlands. A glance at a physiographic map of east Asia, for example, allows an immediate visual comparison between the spacious North China Plain, and the Kanto Plain, Japan's largest and most important level area (Map 7). The latter, the chief population centre, with

*Physical components*

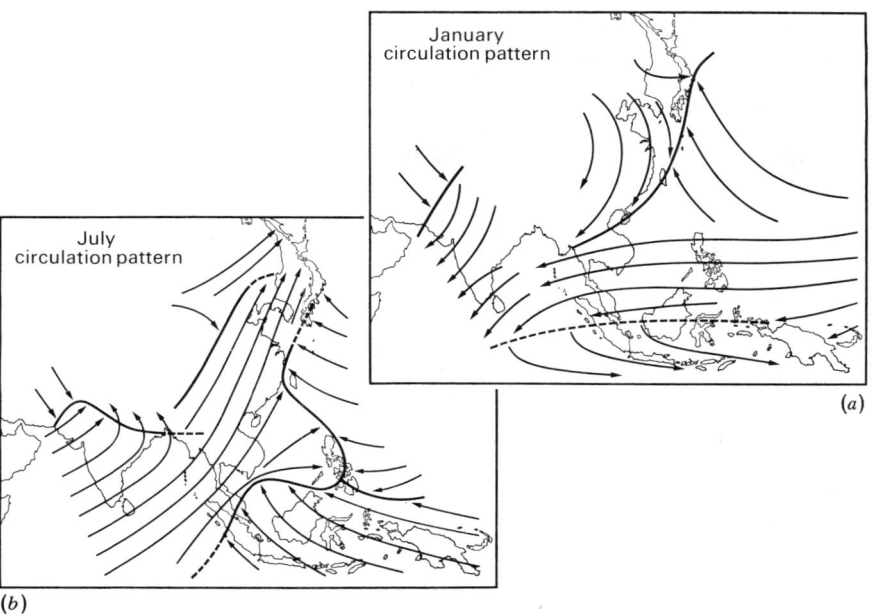

Map 6 Atmospheric circulation, January and July. (*Source*: Trewartha, 1965)
(a) January: Japan lies within the polar-front convergence
(b) July: Japan is influenced both by equatorial southwesterlies and by tropical southeasterlies

around 28 per cent of the world's sixth most populous nation, is only an estimated 32 321 square kilometres in extent and is obviously dwarfed by the vast level areas of northeastern China.[5]

The Kanto Plain, comparative size notwithstanding, is about six times larger than the next broadest plain, the Ishikari-Yufutsu Plain of Hokkaido, and each of the next largest is progressively smaller. These

TABLE 1.2    Total area in slope of less than 15°

| Region | Area ($km^2$) | % of region under 15° | Area ($km^2$) |
|---|---|---|---|
| Honshu | 230 388 | 24.1 | 55 523 |
| Shikoku | 18 754 | 23.2 | 4 350 |
| Kyushu | 41 937 | 26.0 | 10 903 |
| Hokkaido | 78 487 | 26.6 | 20 877 |
| Totals | 369 566 | 24.8 | 91 653 |

*Source*: Adapted from Trewartha, *Japan*, p. 26, Table 1-2. Total area is about that of 1960.

*The urban landscape*

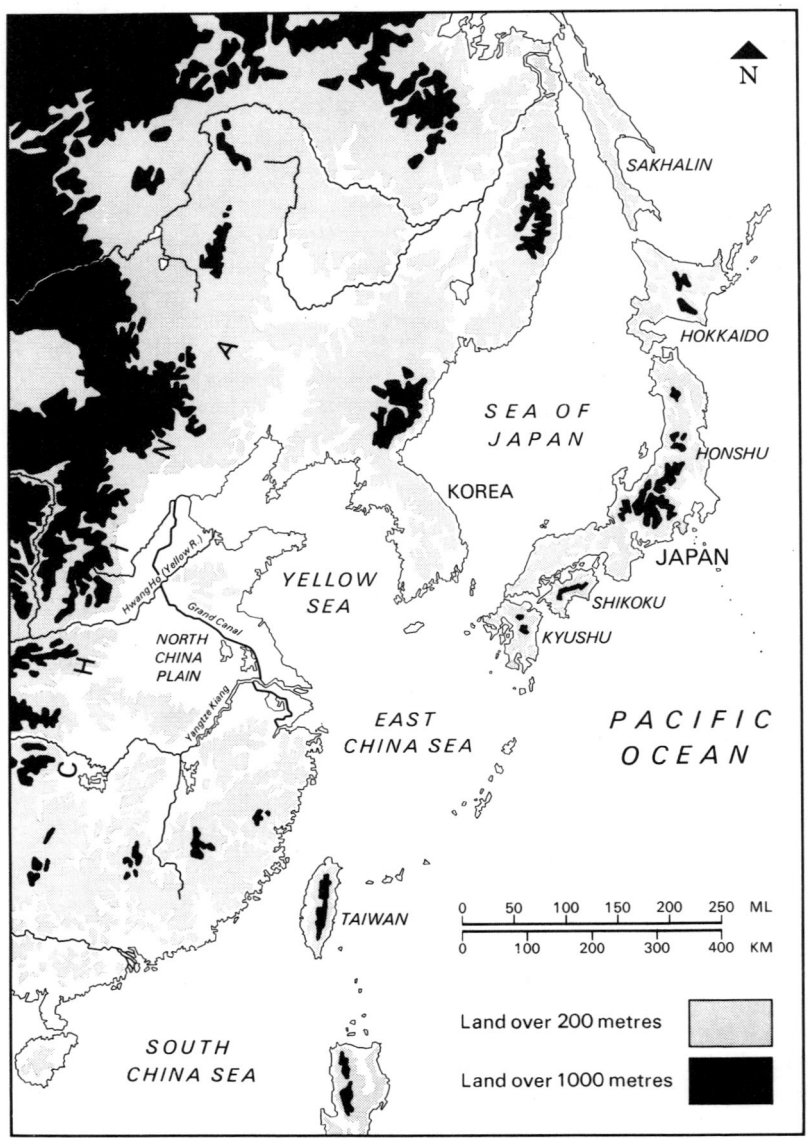

Map 7 Landforms of East Asia, showing distribution of lowland. (*Source*: Cressey, 1955)

include the Niigata or Echigo Plain of northwestern Honshu, the Nobi Plain, associated with the great city of Nagoya and its satellites; the Sendai

Plain of northeastern Honshu, and the Tsukushi Plain, of historic importance in the coal industry of northern Kyushu. The foregoing decrease from about 2 000 to less than 1 300 square kilometres in area, and each is a focal point for settlement and lines of transportation. Even smaller, more fragmented or isolated level lands are frequently of importance as centres of human activity. The Kyoto and Nara Basins, or the deltaic lowlands around Hiroshima, Okayama, Toyama, and other cities, exemplify these.

The main cities of Japan have become the capitals of the major plains and since most of these are along the Pacific shores as far northeast as Sendai, or southwest around the coasts of the Setonaikai to the northwest lowlands of Kyushu, the central core of city life and of most presentday activity lies in an area roughly 885 kilometres long, and inland to various distances depending on the sizes of the plains. Level areas with space for important cities and their hinterlands do exist elsewhere, but generally their cities tend to lag in growth. There are exceptions, especially where raw materials were the initial reason for development, and sometimes such cities have been maintained through the years by their own momentum, even though the original activity is no longer of primary importance.

Within the urban core, particularly between Tokyo and Kobe, competition for space has been so keen that some cities, of which Kobe is an example, have flourished even where appropriate flat land has long been pre-empted. The city has simply expanded up and often over the slopes of what might ordinarily be barrier highlands, and in some cases, rugged land has been bulldozed or flushed away to provide room for new construction, the excess earth being used perhaps for foreshore reclamation, usually for more industry or housing. Most of the level land in the core supports more than one important city, and the rivalry for ascendancy is often acute. In the peripheral areas, the chief cities, themselves places of less consequence in the general urban order, suffer less rivalry, but also less stimulation to attain higher status. Throughout history, and especially in recent years, these cities have tended to lose population to the core, notably of citizens in the more productive age brackets.

## Comparative size

Japan is usually held to be a small country, even, or perhaps especially, by the Japanese themselves. It is small, however, only when compared with the larger of the world's industrially advanced nations. Japan is, for example, larger than all European countries except Sweden, France, and Spain. Moreover, it is latitudinally elongate, so that overland travel between such important centres as the conurbations of northern Kyushu and those of Hokkaido, could mean a continuous journey of about twenty-four hours by the fastest means available today. This would include a slightly more than six-hour dash from Fukuoka to Tokyo on the super-speed line, the *Shinkansen*, the world's fastest railroad; a four-hour ferry trip across the Tsugaru Strait to Hokkaido, as well as the most

*The urban landscape*

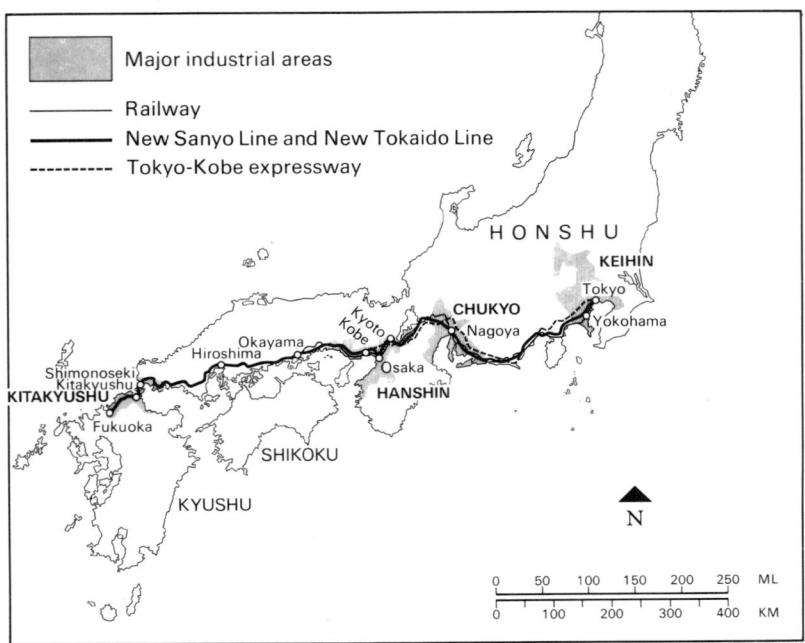

Map 8 Japan: Industrial belt and urban core. (*Source*: Hall, 1976)

advanced diesel- and electric-powered special express trains in use for the remainder. By the 1990s, thanks to extensions of the *Shinkansen* and the completion of the world's longest undersea rail tunnel to Hokkaido, which will eliminate the ferry trip, and perhaps to other technical advances now being tested, it is hoped that this time-distance will be drastically shortened. Present jet airline schedules also sharply reduce distance but even so the flying time required is nearly three hours. Actually, few would attempt unbroken trips of such magnitude by air. The schedules are not usually arranged for this and in addition, as in other places, air travel time is lengthened by the slowness of traffic between cities and airports, despite some relief in recent years provided by new highways.

Cross-island travel is less lengthy, the distances usually being under 320 kilometres. On the other hand, facilities are far more limited and the rough topography greatly reduces average running times of trains along the predominantly single-tracked lines. Automobiles and buses, though now much used for such travel, are also extremely slow because of the scarcity of routes and the narrowness of roads. The aeroplane has provided a partial answer here as well, but the points of contact are limited and scheduled flights are comparatively few.

In short, distance in Japan is still something of a hindrance to move-

## Physical components

ment. There is also the larger consideration that Japan, as a region, is too extensive for many generalisations. Thus, while instability is common, certain regions, such as the Chugoku area of western Honshu, have only minimal tectonic activity. The latitudinal spread, coupled with variegated relief over long stretches, has been especially significant in creating a variety of climates and hence a surprising richness of flora and fauna. Cultural regionalisation, with such manifestations as differences in traditional speech, dress, and other customs, is still felt despite the effects of modern mass movement and communications.

### Natural resources and city development

Raw material placement, a powerful factor in the location and subsequent development of cities in many other advanced industrial nations, has played a role of less importance in Japan, although the meagre resource base in the late nineteenth century provided some temporary impetus. Coal especially was important as a *raison d'être* for the cities of northwestern Kyushu and southern Hokkaido. The general demise of the coal industry, coupled with the fact that Japan's coal is for the most part bituminous material of low grade, has resulted in hardship and consequently population decline in these areas, notwithstanding the modernisation and rationalisation of the industry and the intense demand for coal in the 1960s, particularly by the burgeoning iron and steel industry. Domestic coal, as a general-use fuel (especially in the home) has largely been replaced by petroleum products and natural gas, and the railways have long been converted to diesel or electric power. But coal is still much used for the generation of electricity, usually in conjunction with oil. In 1977 slightly under 44 per cent of the coal produced in Japan was so used, whereas only 31 per cent went to the iron and steel industry.[9] Movements of coal, as might be expected, have been channelled particularly to the core area, and this supply system in modern times has been important in urban-industrial growth.

The occurrence of metals is widespread in Japan and mining is vigorous and modern despite the smallness of many of the operations. Yet as a generator of urban development mining for metals in modern Japan has resulted in remarkably little. A map of the principal mines in about 1971, of which there were nearly fifty locations, reveals only one city of significance, the large and flourishing city of Hitachi in the northeastern coastal outreaches of Kanto, which is based on nearby deposits of sulphuretted ores, including copper.[10] Hitachi is virtually a company town, since the metals which were the basis of its growth gave rise to the prosperity of the huge Hitachi Electric Company, whose products are known the world over.

Petroleum in Japan is produced almost exclusively in Niigata and Akita prefectures, and within the twentieth century this has provided some stimulation, especially for the growth of the two prefectural capitals. But

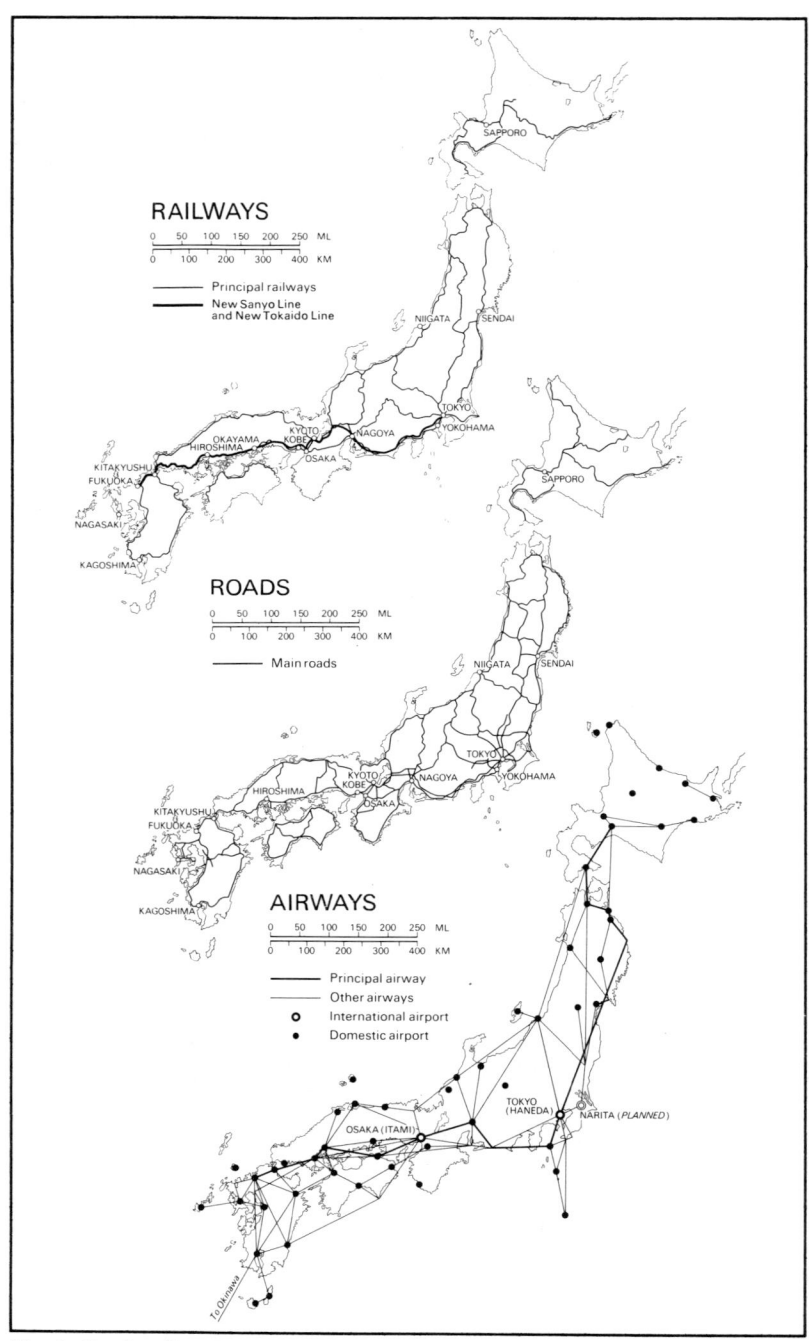

Map 9 Road, rail and communications of Japan. (*Source*: Teikoku, *Complete Atlas of Japan*)

*Physical components*

since more than 99 per cent of the petroleum used in Japan is imported, its local importance in city development is negligible.[11] On the other hand, petroleum importation in the 1960s and early 1970s has led to the strong growth of certain new industrial ports, some of which are striving to become viable cities, with industry based on petroleum and petrochemicals, but mostly these assume significance only if they are located in or near the core. Of some twenty-nine locations of refining centres in 1978, only seven were outside the core and its extensions.[12] Cement, with which Japan is quite well endowed, has probably been of greater importance to cities, especially in southwestern Honshu, northern Kyushu, and in the Kanto area, where cement production is most noteworthy.

The preceding resources have been mentioned in the light of their impact on the locations of cities. Two other important resources are general water resources, and forests. As elsewhere in the world where there are such natural conditions, the mixing off the coasts of Japan of a cold current (the Oyashio, which flows from northern polar seas), with a north-flowing warm current (the Kuroshio, Black or Japan current from equatorial waters), has produced major fishing grounds. In Japan this meeting place is centred off the Chiba Peninsula in the Pacific, its benefits being felt widely around the archipelago, and through the years there have been rich harvests of fish and other aquatic products. In the present day,

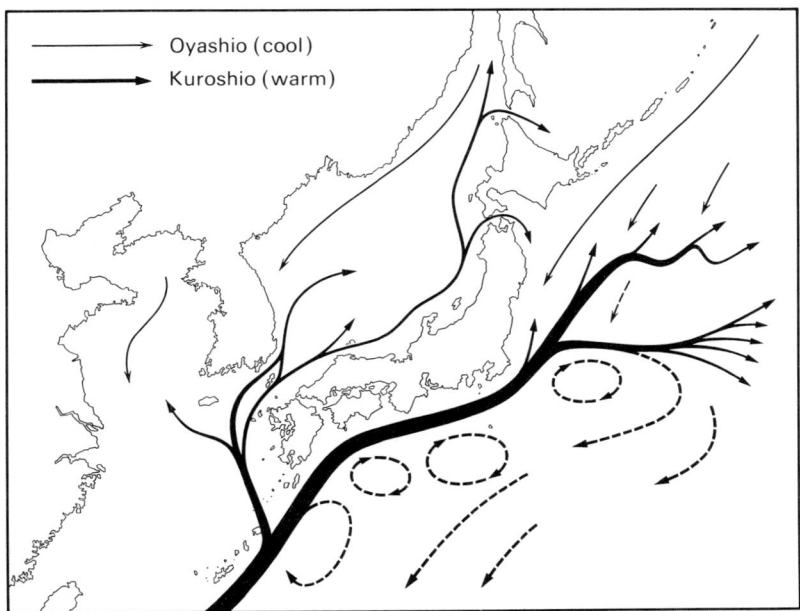

Map 10 Ocean currents in the vicinity of Japan. (*Source*: Trewartha, 1965)

*The urban landscape*

1.3  Oyster fisheries in Hiroshima Bay

because of advancing technology which has helped to contaminate offshore waters and to reduce the bounty, the Japanese have found it necessary to roam the seas of the world to capture an immense variety of marine life; they have become world leaders in this activity, at least in the range of their operations and in the diversity of the catch. Despite dwindling local supplies, however, the immediate offshore waters continue to provide vital resources. Most of the hundreds of Japanese ports play some role in the marine products industry, and fishing and its attendant activities occupy an important segment of the working population. As a way of life this ancient industry is possessed of a special, unmistakably Japanese flavour and expertise, differing in many details from that of other pursuits. The national taste for sea foods is well known. Virtually everything edible is enthusiastically and regularly consumed, and there is a particular fondness for meticulously prepared and attractively served raw fish (*sashimi*), which is both standard fare and a delicacy of unusual piquancy. The cultivation, collection, processing and distribution of marine plants and animals are all practices that have reached an advanced stage of development, as is demonstrated in the daily use of seaweed, dried, canned, frozen and fresh fish in routine food preparation, and also in the thriving export trade in these items. Another culinary delight to exemplify the Japanese mastery over their marine environment is *unagi domburi*, a dish consisting of small (usually cultivated) eels which are broiled and basted with sweetened soy sauce and served on a bed of hot rice. Scientific marine research is also highly advanced, from the university level to such exotic but practical applications as the raising of cultured pearls, techniques for which were originated by Mikimoto Kōkichi beginning in the 1880s.

Nor is the national craving for aquatic foods and other products limited to those taken from the sea. Rivers and streams, ponds and lakes, even inundated paddy fields in season, provide freshwater products. Sport fishing, both freshwater and salt, is universally popular, and there is a lively trade, both domestic and foreign, in locally manufactured fishing gear and other equipment having to do with the sea, or with other water features. The fishing industry as such has provided little major impetus for modern urbanisation, but fishing activity is carried on in all ports, large and small. Fishing has therefore been an important adjunct to the prosperity of many cities.

Natural runoff, as another aspect of water resources, has long been an important source of electrical energy and, when impounded, of irrigation water for the raising of rice and other crops. Rivers and streams are usually too shallow for modern navigation but, as will be explained, have constructed floodplains and fashioned deltas which have been paramount sites for the locations of settlements. Major cities, especially, owe much of their development to such site characteristics.

Forest resources, likewise, have been of inestimable value throughout history for providing building materials, paper, agricultural and industrial implements and raw materials, and fuel. Wood has exerted a powerful aesthetic appeal on the Japanese, not only for building and construction, but also as an artistic medium in the purest sense, as can be seen in the exquisite wood sculpture dating from earliest historical times.

When natural supplies of wood thinned through use, the Japanese evolved and imported methods of replanting and harvesting, and as with the sea, since demand has now outrun the supply, there is enormous supplementation from abroad. In 1977 57.6 per cent of the wood and wood products used in Japan was imported.[13] Orientation around wood and paper has been especially associated with Japanese material culture. Early emphasis on writing and printing prompted the manufacture of paper, an industry which in the present has grown to a point of extraordinary richness and sophistication. Hence, printing and publishing are trades at which the Japanese also excel, not only in their own language or culture, but universally, the high quality of their craft being recognised in the world at large.

As will be mentioned frequently in the following pages, the place of forests in the national culture is shown in many ways, from depiction in the art of all ages to carefully tended public preserves. While magnificent stands of giant *cryptomeria* and other species of natural vegetation are highly esteemed, especially in association with remote religious sites, most hill lands are reafforested and are thus an integral part of the man-made landscape, adding a lush background for the beauties of the agrarian scene and of adjacent seas. Replanted trees are usually of fast-growing varieties of pine, while natural vegetation is made up of a rich assortment of mid-latitude trees and other plants, from conifers in northerly and mountainous areas, to such deciduous types as maple, birch, oak, cherry, and chestnut. Sheltered areas of the west and south are often in semitropical strains such as camphor and palm, and the bamboo, both picturesque and versatile in its usefulness, is prevalent in limited stands throughout western and central Japan. Clumps of trees are often planted around individual farmsteads as windbreaks, which add to the charm of the rural landscape. Tree-studded hills behind farm villages are also the traditional woodlots, often communally owned, which have provided fuel (until broadly replaced in recent years by propane gas) for daily baths and cooking. Obviously, forests in themselves have had little impact on cities, but their importance in the general culture, especially in urban culture, cannot be ignored. Forested lands are also the predominant landscape feature, occupying more than 65 per cent of the total national area in 1975.[14]

## II. Some human components of the urban landscape

### Population and settlement

The paucity of Japan's material assets for major modernisation is so remarkable in virtually every usual category that the *human* resource assumes special importance.

Ethnically, the Japanese are often said to be relatively homogeneous, although their origins are far from clear. Our knowledge of their prehistory is derived mostly from archaeological evidence and from the rather inexact accounts of neighbouring peoples. Notwithstanding, the basic stock is a Mongoloid admixture with early elements of possibly indigenous Caucasoid strains. The physical type of the modern Japanese is so widely varied, however, as to cast doubt on claims of ethnic homogeneity — a moot point — and in any case such attributes are overshadowed in importance by cultural features.

A strong sense of cultural identity among Japanese was established very early, perhaps even before the introduction of Chinese influences, including Buddhism and writing in the sixth century AD. The ensuing historical periods, especially from the ninth century onwards, greatly reinforced this sense of identity and, as mentioned, the imposed isolation between the 1630s and the 1850s succeeded in creating a degree of national spirit that was rare at the time.

Lowland settlement, particularly in the core area, gradually resulted in a series of interacting communities. During the Edo or Tokugawa period (1600—1868), under unusually astute military-political leadership, these were moulded into a coherent (albeit premodern) state, whose population was supported chiefly by agriculture, but was also heavily engaged in secondary and tertiary activities. Chapter 3 outlines the history of city growth and so emphasises these latter aspects of cultural development. The history of population growth is also approached in Chapter 3, and even more briefly below.

The problem of population is particularly acute in Japan because of its lack of natural resources and living space; indeed, the history of concern

## Some human components of the urban landscape

with these matters is long and involved. From about the sixteenth century to at least the 1860s, despite the imprecision of early surveys, a relatively stable situation seems to have existed. The general estimates are that there were between 25 and 30 million Japanese, the figure varying according to agricultural vagaries and conditions of general health. The usual cycle of high births and high deaths, common to premodern cultures, obtained. Families were extended and small, relatively independent marriage relationships were, at best, strongly discouraged. Poverty was often widely felt and infanticide (*mabiki*) was a major check to population expansion. Some of these conditions — notably the extended family, rural poverty, infanticide, and a high birth rate (while the death rate was beginning to decline) — continued after 1868, and thus the population grew quickly as had already happened in the industrialising nations of Europe and North America. Table 1.3 illustrates this clearly.

It is also shown in Table 1.3 that after 1955, and again as in other developed nations, the old cycle of high births and deaths underwent a fundamental change. Since that year the population has grown very moderately at around 1 per cent per year, in sharp contrast to the period immediately after the Second World War, when the demographic cycle was suddenly upset by a new condition of high births and unprecedentedly low deaths. The result was the so-called 'baby boom' of the late 1940s, which saw average increases of more than 3 per cent per annum. Even in the period 1950 to 1955 the growth rate was nearly 2 per cent. But strong measures, of which legalised abortion was the most effective, succeeded in reducing this inflation, and in the 1960s were added such vastly changed circumstances as a radical shift from the extended to the nuclear family, relatively late marriage, modern birth control practices, and the rapidly increasing urbanisation of housing and of life in general.

Yet this new control of population growth has not eradicated the basic problem of increasing population pressure. Furthermore, as the members of the postwar 'baby boom' era have now entered the stage of reproduction, and especially as general affluence continues or increases, the remaining years of this century promise to be a period of new population problems. Demands for housing, always much greater than the means to satisfy them, as well as returning pressures for school and other public facilities, are expected to increase markedly. Moreover, the already large population base of more than 118 million, even at present rates of growth, provides a new increment of well over 1 million persons per year.

### Population density

Figures showing the average number of persons per unit area for a country are usually a poor indicator of living conditions, and this is particularly true of Japan where only a small percentage of the domain is regarded as suitable for most modern activity. The majority live and work on only a fraction of the usable space, so that densities in these areas are among

Table 1.3  General population growth 1875–1970 = quinquennially; 1971–79 = yearly

| Year | Number ('000s) | % increase (quinquennial) | Year | Number ('000s) | % increase (yearly) |
|---|---|---|---|---|---|
| (1872) | 34 806 | — | 1971 | 106 100 | 1.37 |
| 1875 | 35 316 | — | 1972 | 107 595 | 1.41 |
| 1880 | 36 649 | 3.77 | 1973 | 109 104 | 1.40 |
| 1885 | 38 313 | 4.54 | 1974 | 110 573 | 1.35 |
| 1890 | 39 902 | 4.15 | 1975 | 111 940 | 1.24 |
| 1895 | 41 557 | 4.15 | 1976 | 113 089 | 1.03 |
| 1900 | 43 847 | 5.51 | 1977 | 114 154 | 0.94 |
| 1905 | 46 620 | 6.32 | 1978 | 115 174 | 0.89 |
| 1910 | 49 184 | 5.50 | 1979 | 116 133 | 0.83 |
| 1915 | 52 752 | 7.25 | | | |
| 1920 | 55 963 | 6.09 | | | |
| 1925 | 59 737 | 6.74 | | | |
| 1930 | 64 450 | 7.89 | | | |
| 1935 | 69 254 | 7.45 | | | |
| 1940 | 71 933 | 3.87 | | | |
| 1945 | 72 147 | 0.30 | | | |
| 1950* | 84 115 | 16.59 | | | |
| 1955 | 90 077 | 7.09 | | | |
| 1960 | 94 302 | 4.69 | | | |
| 1965 | 99 209 | 5.20 | | | |
| 1970 | 104 665 | 5.50 | | | |

*Sources:* Quinquennial to 1915, *Japan Statistical Yearbook*, 1968, Bureau of Statistics, Office of the Prime Minister, Tokyo, 1969, pp. 10–11; Quinquennial, 1920–70, *1975 Population Census of Japan*, vol. 1, pp. 6–9; Yearly, 1971–78, *Japan Statistical Yearbook*, 1980, Table 7, pp. 12–13.

*All figures after 1950 include Okinawa prefecture.

## Some human components of the urban landscape

1.4 Congested settlement in a small pocket of lowland at the mouth of a valley at Sakashitsu (see also 1.1)

the highest in the world and are increasing with each census. On the other hand, while this demonstrates extreme congestion, this condition is probably in store for most of the world's peoples. As the Japanese themselves have amply demonstrated moreover, crowdedness is an inadequate measure of at least *economic* wellbeing. More will be said in the following pages of the effects of crowding on other aspects of life in Japan.

Tables 1.4, 1.5*a* and 1.5*b* present an added dimension to the matter of population density: at least they reveal in a general way how large percentages of the total population have come to occupy very small percentages of the total area. The highest densities, needless to say, are those of the core area prefectures, though it should be remembered that even these administrative areas contain much rugged land where densities are often minimal.

TABLE 1.4   National population by region, 1980

| Region | Area | | Population | | Density persons/km² |
|---|---|---|---|---|---|
| | (km²) | % | '000s | % | |
| Hokkaido | 78 519 | 21.07 | 5 576 | 4.76 | 71 |
| Honshu | 230 988 | 61.98 | 93 246 | 79.66 | 404 |
| Tohoku | 66 964 | 17.97 | 9 572 | 8.18 | 143 |
| Kanto | 32 343 | 8.68 | 34 894 | 29.81 | 1 079 |
| Chubu | 66 759 | 17.91 | 19 985 | 17.07 | 299 |
| Kinki | 33 055 | 8.87 | 21 209 | 18.12 | 642 |
| Chugoku | 31 867 | 8.55 | 7 586 | 6.48 | 238 |
| Shikoku | 18 800 | 5.04 | 4 163 | 3.56 | 221 |
| Kyushu | 42 130 | 11.30 | 12 965 | 11.08 | 308 |
| Okinawa | 2 249 | 0.61 | 1 107 | 0.94 | 492 |
| Totals | 372 686 | 100.00 | 117 057 | 100.00 | 314 |

Adapted from: *Nihon Kokuseizue* 1981, Table 6−1, p. 66.

**Ways of life**

The city in Japan is the true centre of Japanese culture. Furthermore, as will be explained in Chapter 3, its importance in this light has been felt for centuries. Today, one has only to spend a short time in Japan under routine living conditions to realise that the standards of the city are those for all society. Even by official measurement, most working Japanese are classed as other than rural (Table 1.6), and as is shown in Table 1.7, the bulk of the public, according to the 1975 *Population Census,* are considered to be city-dwellers. The latter figure of 75.9 per cent may exaggerate the actual area that is urban, since the requirements for city status were revised in 1953 to allow the inclusion of much agricultural land within the limits of most cities. But recently the statistician's claim is tending toward reality, as is demonstrated in the proportion of the total population living in DIDs. Between 1970 and 1975, this figure rose by 3.5 per cent, whereas the percentage of the population living in 'All cities' as opposed to those in 'All counties' in Table 1.7, increased by only 1.2 per cent.

In addition to this statistical evidence that the Japanese are predominantly city-oriented, there are other indices in the form of mass-produced goods and services, available throughout the country. Mass communications (known in Japanese as *masu-komi*) show this impressively, especially on the private television networks, whose programmes are rife with the kinds of commercialism, rendered in a pseudo-sophisticated way, that are usually associated with Madison Avenue and the United States.

City living is fairly standard. That is, the average routine of life in one

TABLE 1.5a  Densely inhabited district* populations encircling the largest cities, 1975

| Cities and radii | | DID population | | | | Total populations | | | | % DID of total | |
|---|---|---|---|---|---|---|---|---|---|---|---|
| | | 1975 | 1970 | Change | % | 1975 | 1970 | Change | % | 1975 | 1970 |
| Totals | 50 km | 38 468 679 | 33 278 320 | 5 190 359 | 15.6 | 47 062 236 | 42 385 118 | 4 667 118 | 11.0 | 81.7 | 78.5 |
| Tokyo | 50 km | 21 347 867 | 18 236 104 | 3 111 763 | 17.1 | 24 760 701 | 21 970 993 | 2 789 708 | 12.7 | 86.2 | 83.0 |
| Osaka | 50 km | 12 879 197 | 11 467 526 | 1 411 671 | 12.3 | 14 871 590 | 13 639 832 | 1 231 758 | 9.0 | 86.6 | 84.1 |
| Nagoya | 50 km | 4 241 615 | 3 574 690 | 666 925 | 18.7 | 7 429 945 | 6 774 293 | 655 652 | 9.7 | 57.1 | 52.8 |
| Tokyo | 70 km | 22 227 811 | 18 942 113 | 3 285 698 | 17.4 | 27 525 701 | 24 495 633 | 3 030 068 | 12.4 | 80.8 | 77.3 |

TABLE 1.5b  Areas of densely inhabited districts encircling the largest cities, 1975

| Cities and radii | | Areas (unit: km$^2$) of DID and % | | | | Total areas | | % DID of total | |
|---|---|---|---|---|---|---|---|---|---|
| | | 1975 | 1970 | Change | % | 1975 | | 1975 | 1970 |
| Totals | 50 km | 4 065.3 | 3 041.6 | 1 023.7 | 33.7 | 22 266.3 | | 18.3 | 13.7 |
| Tokyo | 50 km | 2 190.9 | 1 708.4 | 482.5 | 28.2 | 7 609.2 | | 28.8 | 22.5 |
| Osaka | 50 km | 1 243.8 | 885.3 | 358.5 | 40.5 | 7 349.3 | | 16.9 | 12.0 |
| Nagoya | 50 km | 630.6 | 447.9 | 182.7 | 40.8 | 7 307.8 | | 8.6 | 6.1 |
| Tokyo | 70 km | 2 342.9 | 1 812.8 | 530.1 | 29.2 | 13 701.7 | | 17.1 | 13.2 |

Source: adapted from *1975 Population Census of Japan*, Densely Inhabited Districts, p. 11, Tokyo, 1977.

*Densely inhabited districts (DID) were instituted in the general census of 1960 as a more precise means to measure urban population. In 1975 DIDs contained 57.0% of the total population, compared to 53.5% in 1970, 48.1% in 1965, and 43.7% in 1960, according to the *1975 Population Census of Japan*, vol. 1, pp. 22–3, Tokyo, 1977.

TABLE 1.6  Employed persons by types of industry, 1960–75

| Year and % | Totals | Types of industry (unit: 10 000 persons) | | | | | | | | | | | | |
|---|---|---|---|---|---|---|---|---|---|---|---|---|---|
| | | (1) | (2) | (3) | (4) | (5) | (6) | (7) | (8) | (9) | (10) | (11) | (12) | (13) |
| 1960 | 43 691 | 13 121 | 439 | 676 | 537 | 2 674 | 9 553 | 6 920 | 701 | 82 | 2 207 | 233 | 5 211 | 1 328 |
| % | 100.00 | 30.0 | 1.0 | 1.5 | 1.2 | 6.1 | 21.9 | 15.8 | 1.6 | 0.2 | 5.1 | 0.5 | 11.9 | 3.1 |
| 1965 | 47 610 | 10 867 | 262 | 603 | 332 | 3 376 | 11 676 | 8 486 | 951 | 201 | 2 849 | 263 | 6 227 | 1 489 |
| % | 100.00 | 22.8 | 0.6 | 1.3 | 0.7 | 7.1 | 24.5 | 17.8 | 2.0 | 0.4 | 6.0 | 0.6 | 13.1 | 3.1 |
| 1970 | 52 235 | 9 334 | 206 | 535 | 216 | 3 929 | 13 682 | 10 060 | 1 104 | 273 | 3 214 | 287 | 7 635 | 1 720 |
| % | 100.00 | 17.9 | 0.4 | 1.0 | 0.4 | 7.5 | 26.2 | 19.3 | 2.1 | 0.5 | 6.2 | 0.6 | 14.6 | 3.3 |
| 1975 | 53 141* | 6 700 | 179 | 475 | 132 | 4 729 | 13 236 | 11 381 | 1 383 | 372 | 3 365 | 321 | 8 741 | 1 959 |
| % | 100.00 | 12.7 | 0.3 | 0.9 | 0.2 | 8.9 | 25.0 | 21.5 | 2.6 | 0.7 | 6.4 | 0.6 | 16.5 | 3.7 |

*Source: Japan Statistical Yearbook, 1980,* pp. 56–7. Types of industry: (1) agriculture; (2) forestry and related; (3) fishing and related; (4) mining; (5) construction; (6) manufacturing; (7) wholesale and retail trade; (8) finance and insurance; (9) real estate; (10) transportation; (11) electricity, gas, water; (12) services; (13) government.

*The 1975 total figure appears as above in the source, while the categories actually total 52,973, a difference of 168. Percentage figures for 1975 are therefore approximate in a few cases. There are small differences in all other totals as well (also, as in the source) but never enough to change the percentage figures shown.

TABLE 1.7    Percentages of population in cities and counties, 1920−75

| Year | Size of city | In all cities | In all counties |
|---|---|---|---|
| 1975 | | 75.9 | 24.1 |
| | Over 50,000 | 67.3 | − |
| | Under 50,000 | 8.6 | − |
| 1970 | | 74.7 | 25.3 |
| | Over 50,000 | 65.7 | − |
| | Under 50,000 | 8.9 | − |
| 1960 | | 63.3 | 36.7 |
| 1955 | | 56.1 | 43.9 |
| 1950 | | 37.3 | 62.7 |
| 1947 | | 33.1 | 66.9 |
| 1945 | | 27.8 | 72.2 |
| 1940 | | 37.7 | 62.3 |
| 1935 | | 32.7 | 67.3 |
| 1930 | | 24.0 | 76.0 |
| 1925 | | 21.6 | 78.4 |
| 1920 | | 18.0 | 82.0 |

Source: *1975 Population Census of Japan*, vol. 1, p. 2.

city differs little from that in another. In the eyes of the Japanese, however, there is often more prestige attached to the larger metropolitan centres which have an unquestioned edge in amenities, and probably a greater wealth of goods and services. But many feel that the price is higher, not only in the raw cost of living, but also in the quality of life which tends to appeal most to the young and relatively footloose, for there is more chance for excitement and especially for anonymity than in the smaller communities. The larger cities characteristically have a high male sex ratio and a median age in the more productive years.[15]

For the purposes of this account, however, a typically urban Japanese would not necessarily live within one of the half-dozen or so giant cities, although he probably would be under a strong urban influence, and certainly his life would be especially influenced by Tokyo. The typical urban Japanese might not live within the limits of a major city at all, but might simply commute from a satellite settlement, often called a city for administrative purposes. The proliferation of such agglomerations within the past decade has been an outstanding feature of contemporary urban population growth, not only in the interstices of the established urban core but also generally. Even those who ostensibly live on farms are mainly dependent on non-farm incomes, and most of this is accomplished by commuting to nearby cities, with farm labour shortages being met by increasing mechanisation and the participation of entire households only in times of heaviest need in the agricultural cycle. Such practice, known as 'part-time farming', is the principal technique for dealing with the increasing dearth of labour in Japanese agriculture.

## Changes in types of settlement

The number of cities and towns has sharply increased since 1955, and the number of villages has dwindled to less than 1 000, mainly because of administrative inducement through a Government programme of amalgamation in the 1950s, which encouraged the regrouping of towns and cities into larger units (Table 1.8). In 1978 there were 645 cities, most of which had populations of at least 50 000.[16]

TABLE 1.8    Changes in numbers of types of settlement, 1883–1978

| Year | Cities (A) | Towns | Villages | Total (B) | Ratio: A/B |
|---|---|---|---|---|---|
| 1883 | 19 | 12 194 | 59 284 | 71 497 | 0.03 |
| 1890 | 40* | 15 732† | – | 15 772 | 0.25 |
| 1895 | 41 | 15 804† | – | 15 845 | 0.26 |
| 1900 | 53 | 14 044† | – | 14 097 | 0.38 |
| 1905 | 55 | 13 428† | – | 13 483 | 0.41 |
| 1910 | 61 | 1 140 | 10 751 | 11 952 | 0.51 |
| 1915 | 65 | 1 271 | 10 534 | 11 870 | 0.55 |
| 1920‡ | 77 | 1 344 | 10 794 | 12 215 | 0.63 |
| 1925 | 101 | 1 504 | 10 440 | 12 045 | 0.84 |
| 1930 | 109 | 1 702 | 9 980 | 11 791 | 0.92 |
| 1935 | 129 | 1 702 | 9 721 | 11 552 | 1.12 |
| 1940 | 160 | 1 754 | 9 325 | 11 239 | 1.42 |
| 1945 | 205 | 1 797 | 8 518 | 10 520 | 1.95 |
| 1950 | 235 | 1 862 | 8 346 | 10 443 | 2.25 |
| 1955 | 488 | 1 833 | 2 885 | 5 206 | 9.37 |
| 1960 | 555 | 1 922 | 1 049 | 3 526 | 15.74 |
| 1965 | 560 | 2 005 | 827 | 3 392 | 16.51 |
| 1970 | 578 | 2 013 | 684 | 3 275 | 17.65 |
| 1975 | 643 | 1 974 | 640 | 3 257 | 19.74 |
| 1978 | 645 | 1 985 | 626 | 3 256 | 19.81 |

*Sources:* 1883–1965, *Toshi Mondai* (Municipal Problems), vol. 60, July 1969, pp. 91–3; 1970–1978, *Zenkoku Shi, Chō, Son Yōran 53* (National City, Town and Village Yearbook), Tokyo, 1978, p. 3.

*Cities were first incorporated in 1889.
†Includes villages.
‡First modern national census.

It would, of course, be misleading to imply that such development has been uniform in Japan, for urbanisation has followed a pattern of increasing nodalisation familiar elsewhere in the world. The great conurbations have, particularly since 1960, rapidly grown towards one another to form an almost continuous urban region, from a broad area around Tokyo in the east to well beyond Kobe in the west. The 'Tokaido Megalopolis' is a term much used today, popularly as well as professionally. 'Tokaido' is taken from the most famous and important of the feudal highways, with which the region corresponds roughly in location; in the Edo period it was

*Some human components of the urban landscape*

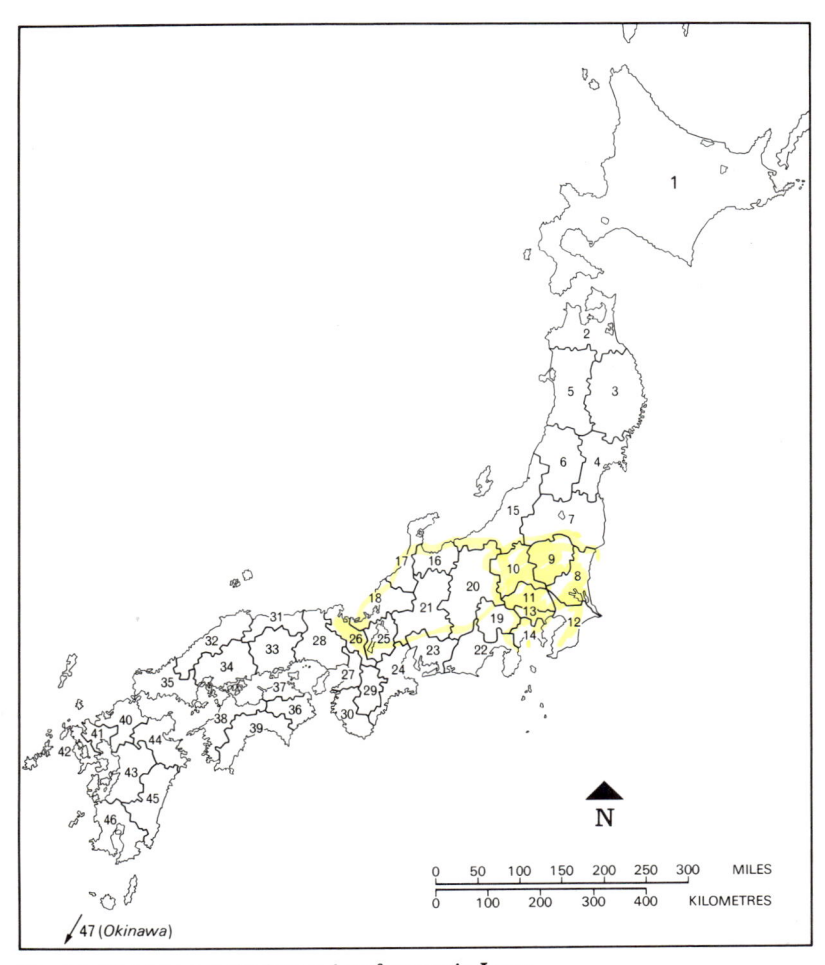

Map 11  Districts, subdistricts and prefectures in Japan

| | | | |
|---|---|---|---|
| 1 HOKKAIDO | KANTO | CHUBU | CHUBU (Tosan) |
| TOHOKU | 8 Ibaraki | (Hokuriku) | 19 Yamanashi |
| 2 Aomori | 9 Tochigi | 15 Niigata | 20 Nagano |
| 3 Iwate | 10 Gumma | 16 Toyama | 21 Gifu |
| 4 Miyagi | 11 Saitama | 17 Ishikawa | CHUBU (Tokai) |
| 5 Akita | 12 Chiba | 18 Fukui | 22 Shizuoka |
| 6 Yamagata | 13 Tokyo | | 23 Aichi |
| 7 Fukushima | 14 Kanagawa | | |
| KINKI | CHUGOKU | SHIKOKU | KYUSHU |
| 24 Mie | 31 Tottori | 36 Tokushima | 40 Fukuoka |
| 25 Shiga | 32 Shimane | 37 Kagawa | 41 Saga |
| 26 Kyoto | 33 Okayama | 38 Ehime | 42 Nagasaki |
| 27 Osaka | 34 Hiroshima | 39 Kochi | 43 Kumamoto |
| 28 Hyogo | 35 Yamaguchi | | 44 Oita |
| 29 Nara | | | 45 Miyazaki |
| 30 Wakayama | | | 46 Kagoshima |
| | | | 47 Okinawa |

the chief link between the military capital of Edo (modern Tokyo) and the imperial capital of Kyoto, and as such it will be discussed in Chapter 3.

## Some political aspects of contemporary urbanisation

In offsetting imbalances in the present urban pattern, resulting chiefly from the growth in modern times of channels of communication that have been fostered mainly by economic forces, the role of politics is most important. The very political organisation of the nation is, in itself, heavily responsible for the maintenance of cities of considerable vitality outside the core area.

Japan proper (excluding Okinawa) is administratively divided into forty-two prefectures (*ken*), the two urban prefectures (*fu*) of Osaka and Kyoto, the metropolitan prefecture (*to*) of Tokyo, and the entire island (*do*) of Hokkaido (Map 11). Each has a governor (now appointed from Tokyo), a deliberative assembly, and certain powers of its own, although there is little of the kind of local autonomy enjoyed, for example, by the states of the United States. This situation has obtained since the 1890s with little change except for minor boundary shifts, although the elevation of Tokyo to *to* status, and of Osaka and Kyoto to *fu*, is relatively recent. There have been, moreover, occasional demands for the reorganisation of prefectures, especially where administrative functions of the great cities are impeded by multiple jurisdictions. This is especially true around Tokyo, Nagoya, and Osaka.

On the local level there are cities (*shi*), towns (*chō*), and villages (*son*) — again with legislative mechanisms and administrative leaders, all of whom are engaged in furthering the positions of their communities. School children learn the entire arrangement by repeating the Chinese readings of these terms in doggerel fashion, *to, dō, fu, ken, shi, cho, son*. The Japanese readings for towns (*machi*) and villages (*mura*), may also be used freely. Changes on the local level have been constant, as can be seen in Table 1.8.

Often parallel with the names of official administrative units is a series of titles for relatively indefinite regions or districts. Usually, these names have historical connotation and as such, are generally understood and much used. But to the outsider — even at times to a Japanese — they are confusing. Hence, there is often an inconsistent application of terms for a single area (for example, *Kinki, Kansai* or *Kinai* for the area around Osaka and Kobe), although each may refer to a slightly different area, or may be used in a special context. Currency of expression is also a factor in their use.

More precise, but still unofficial, are the groupings of prefectures into regions (Map 11), or the telescoping of Chinese character readings for urban districts, such as *Keihin* for Tokyo–Yokohama; *Keihanshin*, for Kyoto–Osaka–Kobe, or simply *Hanshin* with Kyoto omitted.

Returning to cities as such, each is also involved in public relations to present a favourable 'image' to the outside world. This can be seen in the readily available tourist-oriented literature, or in the maintenance of offices for this purpose in Tokyo, or in the participation of the city in sister-city relationships with other of the world's cities that are considered appropriate partners in size and function. Hence, each administrative unit, at least to the city level, is also engaged in lobbying for governmental support which is from time to time extended and which has generally been an important factor in its development. An elaborate system of 'equalisation grants' is also in effect. These are allotments or adjustments of public funds from the central Government to the prefectures to bolster local finances, often for cities that have tended to lag in growth.

Some measure of the advantages of location in the core and its extensions can be seen in the funding of new construction by various agencies. An analysis by prefecture in 1978 shows that the twelve core prefectures of Saitama, Chiba, Tokyo, Kanagawa, Gifu, Shizuoka, Aichi, Mie, Shiga, Kyoto, Osaka, and Hyogo, received 51 per cent of the total construction funds. And, if the following fifteen are considered extensions of the core, the figure becomes 75 per cent. The additional prefectures are: Miyagi, Fukushima, Ibaraki, Tochigi, Gumma, Nara, Wakayama, Okayama, Hiroshima, Yamaguchi, Kagawa, Ehime, Fukuoka, Saga, and Kumamoto.[17]

# 2
# The agrarian landscape

To Westerners, especially to Americans familiar with rural landscapes that show few signs of human occupation or reflect wasteful land management practices, the Japanese agricultural scene at the height of the growing season, with its precise juxtaposition of useful products systematically arranged and devoid of weeds and other extraneous matter, frequently has come as a shock. So impressive is this landscape that even trained observers have been liable to lose sight of the importance of the far less attractive or logically arranged *urban* environment. Of course, even in Japan there are loosely managed agrarian landscapes and semi-wild areas where agriculture is rare or non-existent, but for most of the cultivable parts of Old Japan the meticulous beauty of the countryside in summer is without doubt one of the chief hallmarks of the culture. In Hokkaido, because of the recency of settlement, certain agrarian landscapes may be more akin to those of the eastern North American outlands, but the more productive areas even of this frontier-like region reflect most strongly the influences of the intensive agricultural practices associated with Old Japan.

The emphasis on irrigated rice culture since prehistoric times is well established. Indeed, judging by the present Government's policy of agricultural subsidies, this emphasis has persisted down to the present. Within the past decade or so, however, a change of major significance has occurred. For the first time in history, despite a waning supply of labour and an expanding population base, the traditional condition of chronic underproduction had been transformed into one of glut. To meet this unprecedented situation the Government has published a set of guidelines for agricultural policy in the 1980s which are intended to 'ensure a smooth changeover from rice cultivation in order to increase self-sufficiency in other crops'.[1]

## Background to the present

The components of the contemporary agrarian landscape are too diverse or complex to be presented adequately in a brief survey, for they comprise a vast array of physical, economic and cultural factors whose origins may be obscure. As has already been suggested, moreover, the present period may well mark a turning to a new kind of rural scene, perhaps towards a

*Background to the present*

2.1 Factories in fields: spring farming activity in conjunction with light manufacturing

landscape of large holdings maintained by a variety of machines and few people, for although agriculture remains at least an avocation for nearly half the working population, average income is now mainly from work outside the field. On the other hand, while new laws whose effects have hardly begun to be realised may eventually reshape the landscape, the agrarian scene in summer has superficially lost little of its traditional flavour, except where urban expansion and its offshoots have managed to encroach on it. A brief examination of some of the basic factors that have contributed to the present configuration may provide needed perspective.

In the first place, the lowlands which formed the setting of the original agrarian culture were generally composed of azonal soils of relatively recent origin which were constantly resuscitated by periodic flooding. These soils are naturally the most productive, but on the whole the soils of Japan are of inferior quality despite their mainly volcanic origin, as they are inclined to have acidic rather than basic properties. In very early times, when there were comparatively few people, natural refreshment through seasonal deposition was probably sufficient to sustain normal production, although frequently there must have been nagging shortages growing out of local interruptions in production and especially in distribution.

The rice cycle around which all agriculture is based, from earliest times and in some measure despite contemporary technological change, was patterned by such natural conditions as relatively regular fluctuations of temperature and precipitation. Gentle, cyclonic rainfall in mid-spring, for example, softened and allowed the rather easy working of the surface which was helpful in the creation of rice seedbeds but was not so prolonged as to interrupt the harvesting of dry-field (winter) crops; and the nearly one month of dry, clear weather that followed permitted, simultaneously, the completion of the winter harvest, the preparation of the same fields for paddy production by tilling, smoothing of the surface and flooding, and finally the transplanting of the seedlings from seedbeds to the soft, water-covered surfaces of the plots. Then there was usually about one month of comparatively steady, warm and soaking rainfall known as the *baiu* or season of monsoon rains, during which the seedlings were carefully nurtured into healthy plants. After this, from about mid-August until the crop was harvested in mid-autumn, except for a period of typhoon rainfall, heaviest in September, which brought danger of damaging floods, the skies were generally bright and the temperatures were warm both day and night. Fields were drained at the beginning of this cycle and the grain was allowed to ripen and turn from shades of green to rich gold, ultimately becoming so heavy that by harvest time the landscape was dominated by yellow masses of stalks, forced by the weight of the grain to lie, bent and drooping, close to the ground. The autumn harvest, ordinarily a time of feverish activity, involved the cutting and drying of the plants and the separation and processing of the kernels which were finally bagged and made ready for sale. The inevitable transformation of the grain from brown to white entailed the removal of the inner husk by stone grinding. This operation is now usually performed in a rice mill as a semi-urban function, but water-powered mills were common in the past.

Thus the nature of the surface, coupled with a climatic cycle that normally provided moisture and warmth when needed and dry conditions for ripening, were basic natural advantages which the early Japanese exploited in order to overcome such disadvantages as the dearth of level and cultivable land, the depletion of soil fertility, damaging variations in the climatic cycle, and the effects of blight and insects. Much of the very recent success of Japanese agriculture can be attributed to the eradication through modern technology of these and attendant ills, but the struggle for environmental control began almost before history and accomplished much long before the advent of modernisation.

The means by which the early Japanese tried to overcome some of the shortcomings of their environment go far to reveal their response to physical challenges that could have been overwhelming; they also shed much light on the nature of the culture itself. The success of the Japanese in creating a viable agricultural base against very heavy odds is an inspiring chapter in man's relations with his physical environment and provides eloquent support for the sentiment that human motivation can virtually nullify environmental obstacles.

The physical resources for agriculture in Japan — with its inordinately confined and scarce level areas, its thin and basically unproductive soils, and its climatic and other natural uncertainties — may seem to those of other cultural orientations to have been singularly unpromising for a culture based on rice production. On the other hand, several factors help to account for the large measure of success that has been achieved. Perhaps the most important of these is the early and sustained concentration on the product itself.

## Elements of agricultural evolution

The prehistoric Japanese are known to have developed a predominantly agrarian culture with strong influences from south China and especially with direct assistance from Korea. Within the family and clan, daily life patterns must have been based on an agricultural cycle built around a deep appreciation of, or even a reverence for, the nutritional and other qualities of irrigated rice, whose production was the leading agrarian occupation from as early as the second century BC, at least for Old Japan from Kyushu through the Kinai and perhaps even as far eastward as the Kanto regions of Honshu. Such was the concentration on this crop that other forms of agriculture, including the cultivation of natural terraces, tablelands and uplands, or generally of places that would have been naturally unsuited to irrigated rice, were pursued only supplementally.[2] For example, upland agriculture was far less intensive and produced either less desirable food crops such as millet, or green manure and compost for the enrichment of paddy lands. Grasslands were used similarly or as pasture for lowland draught animals. Dairying and transhumance as practised in southern Europe were unknown. Forests early supplied nuts, and later fruit, as well as fuel and materials for building and implements. As time went on, forested areas were utilised for the storage of irrigation water, and much later for flood control and commercial tree production. Most agriculture outside the level areas used directly for rice was traditionally designed to supplement the basic activity of raising paddy.

Imported tools and techniques, mostly from Korea but often originating in China, gradually helped to improve agricultural output and to enlarge the cultivable area. Even the earliest varieties of rice used by the Japanese, for example, are considered by certain archaeologists to have been related to those of the Yangtse lowlands of China, with the Koreans perhaps as donors.[3] Crude wooden implements which had probably been adequate to work the soft surface material of the natural marshlands used originally for rice cultivation seem to have been replaced after about the year AD 300 by iron tools, which allowed field sizes to be expanded and paddy culture to be extended into previously unusable areas. At about the same time new techniques for constructing embankments, irrigation canals and ponds were introduced, and Korean technicians were often employed to super-

## The agrarian landscape

vise and demonstrate new skills which later were disseminated from the Kinai lowlands, where this activity was most prominent, to other areas.

By the seventh century these innovations were sufficient to have converted the outer edges of many alluvial fans into level irrigated plots fed by springs typical of such locations. By the year 646 a revolutionary system of cadastral survey had been imported from T'ang China as part of the so-called *Taika Reforms*. The latter method of rectangular survey, known in Japan as the *jō-ri* system, and a kind of harbinger of the nineteenth century 'range-and-township' system of the western United States, resulted in the widespread commitment of agricultural and other lands to reorganisation on a grid plan. This system, although outmoded since the Heian period and inappropriate to the physical configuration of Japan, is still evident in the regions where it was once prominent, notably in the Kinai and Setouchi districts of central and western Honshu, and also in parts of Kyushu and eastern Honshu. Its effects can be recognised in the rectilinear outlines of fields, as well as in historical maps and city plans. The main surviving example of a city that reflects this influence is Kyoto, whose rectangular street plan is derived from such a background.

In the period of agricultural estates (*shōen*) — roughly from the ninth century to the fifteenth — much new paddy land was created from such areas as the backmarshes of valley floors, partly by using a system of multipurpose dead-end ditches. Parts of the important Tsukushi Plain of northern Kyushu, for example, owe their early prominence to this period of development and to this day exhibit remnants of the ancient techniques. After the eighth century valley floors were generally rid of marshy tracts, thanks to new means of water control introduced mainly from Korea. Virtually all important estates of the time were founded on formerly marshy areas.

Also in the *Shōen* period gravity-fed irrigation systems using elevated ponds and ditches to transport the water were widely developed, especially in the Setouchi district of western Honshu and in northern Shikoku, where average rainfall is barely sufficient for rice culture. In other areas, such as on the Osaka Plain or on islands of the Setonaikai, the use of wells for irrigation became common at this time, and there was important terracing of rugged mountain slopes in Shikoku and Kyushu. The latter landscapes are among the most spectacular of the present agrarian scene.

The *Sengoku*, or period of the 'Warring States', from the late fifteenth century until the beginning of the seventeenth, was one of marked development of the agricultural landscape, despite the fairly continuous disruption of normal activity. Techniques of terracing reached a high point and the agrarian domain was greatly widened. The construction of irrigation canal systems was furthered, allowing the conversion to paddy fields of many steeply sloping alluvial fans, of which the Kofu Basin of modern Yamanashi prefecture is the foremost example. The draining and reclamation of far-flung areas of swamp, as on the Niigata Plain of northwestern Honshu, are noteworthy achievements of this period and came about

*Elements of agricultural evolution*

2.2*a* Fertilised seedlings being prepared for germination in boxes
  *b* Seedlings being transplanted by machine after about six weeks. Note pontoons on machine

*The agrarian landscape*

because of the closer control of peasant labour by local warlords. Because of internecine strife and a dearth of funds few really large-scale projects were undertaken; nevertheless, the period saw much progress and was, in a sense, a time of preparation for the efflorescence of agriculture during the age that followed.

### Edo period developments

The growth of agriculture was especially remarkable in the early part of the Tokugawa (Edo) period, between the years 1600 and 1720, when the energies of the Japanese were for the first time concentrated under unified direction and when the suppression of internecine wars freed the peasantry from the disruptions which had previously been a regular part of their lives. Furthermore, the weaknesses of the feudal government, especially the economic deterioration that marked the latter part of the period, had not yet become crucial and there was a long stretch of vigorous growth and resulting prosperity.

In this span of years, most of the alluvial plains of Old Japan came to be wholly devoted to paddy culture; bay-heads, natural ponds and many lakes were reclaimed and brought into production, and there was particular expansion of wet rice lands in Tohoku. Indeed it was during this period that such areas as the Echigo (Niigata) Plain grew in importance, and the traffic in grain from that area to Kinai and Edo was integrated into the whole structure of Edo period economy, which was based predominantly on rice.

The fundamental economic structure of the regime was thus of especial importance in furthering agriculture, for a barter system in rice prevailed, rather than a cash economy such as developed in the latter half of the Edo period. Each local lord was accorded status commensurate with the productivity of his lands. Hence there was considerable impetus to improve production and to enlarge the acreage through conversion of marginal lands. Also technicians became more available for such endeavour because of shogunal decrees against the building of castles, which might otherwise have pre-empted their services. Necessary capital was provided by the merchant class, which was rising in economic power, while the labour force was more readily available for large-scale projects; political control was becoming more and more autocratic and increasing emphasis was being placed on such traditional virtues as austerity, obedience, loyalty to ones' superiors, and cheerfulness in the face of dire circumstances.

Projects such as those which had been carried out in the Sengoku period for converting alluvial fans and deltas through the construction of long and intricate systems of drainage were now vastly increased in number and scale. Some of these were surprisingly sophisticated from an engineering standpoint. This is especially true of the volcanic flank areas of modern Nagano and Yamanashi prefectures, where canals were dug sometimes as long as 30 kilometres and were tortuously tunnelled through or around highlands. As a result, such areas came to have unusually heavy rural population densities, even in intermontane parts where flat land was generally deficient.

Pond and lake reclamation projects were also numerous. By the end of this period over 70 per cent of the present total of reclaimed natural pond and lake bottom land in central and western Japan had been converted to usable paddy acreage. It became more or less the fashion for certain large, urban financial interests, especially of Osaka, Kyoto, Sakai, and of Edo itself, to support such construction.

The greatest gains in reclamation, however, were made in bay-head areas and again by the end of this time more than 70 per cent of the reclaimable foreshores of central and western Japan had been transformed into paddy land. By about 1720 the following percentages of the reclaimable parts of these areas had been converted: Ise Bay, 70; Osaka Bay, 68; Kojima Bay, 50 (Okayama prefecture); the Ariake Sea of northwestern Kyushu, 88; and Yatsushiro Bay, south of the Ariake area, 40.

Foreshore reclamation in general, however, was beset by engineering problems, especially at the outer fringes of deltas occupied by numerous distributaries liable to inundation by brackish water. This meant that each plot had to be diked with particular care, the height of the embankment depending on the distance from the main distributary as well as on the maximum floodwater level at its mouth, its length depending on the extent of shoal area being reclaimed. Most representative of this kind of reclamation is the southern part of the Nobi Plain in the Chukyo Region at the head of Ise Bay, where there are gigantic spaces containing settlements as well as growing areas, each circled by elaborate embankments and often presenting from the air a honeycomb effect. Three-fifths of these polders (*waju*) had been completed by early Tokugawa times and of some seventy extant today the largest is about 50 kilometres in circumference. The whole region is prone to damaging floods, the most recent of which was the disastrous Ise Bay typhoon of September 1959; but flood danger has been reduced here in modern times, especially by technological devices including pumps and larger, more secure embankments.

The conversion of marsh and other lands in Tohoku grew out of much earlier settlement including occupation by military pioneers who had migrated here at the time of Hideyoshi in the late sixteenth century. In early Tokugawa, as stated, the agricultural development of this area was accelerated, especially by certain local lords, of whom the famous Date family of Sendai were noteworthy. However, developments were most vigorous along the coasts of the Sea of Japan, where coalitions of merchants and *daimyō* fostered the growth of large-scale rice exporting to the centres of the country. The expansion of paddy land in the Niigata, Shonai and Akita Plains was likewise supported by merchant capital from Kinai as well as by important local figures such as the Homma family of Yamagata who imported labour from Osaka for Shōnai Plain reclamation. By 1720 nearly all the lands of the Sea of Japan littoral had become converted and a pattern of fields virtually identical with that of the present had been established. Further north in the lowlands near the Tsugaru Strait between northern Honshu and Hokkaido, peat bogs had been removed and paddy fields created. Coastwise shipping had also become extremely active despite limitations by the regime on the sizes of vessels.

### Late Tokugawa period developments

As in the years before 1600, the remainder of the Edo period after 1720 is marked by progress in agriculture offset by some stagnation or decline, although this was now caused by economic rather than military circumstances.

*Daimyō* continued and even increased their efforts to enlarge paddy acreage. Bay-head reclamation went on apace in such areas as the Ise and Kojima bays and on the foreshores of the Ariake Sea. Lands in such locations as volcanic flanks, diluvial terraces and tablelands, or generally in places that had been considered of extremely limited usefulness were turned more vigorously at this time into *shinden* (new paddy fields), which were patterned in long rectangular strips, with the narrow ends facing a road, along which, at right angles to the main direction of the fields, houses and other buildings were arranged in villages of the *Strassendorf* variety. This kind of development was particularly common on the Musashino upland of western Kanto, but was known widely elsewhere. Large outlays of capital by the Maeda lords of present-day Toyama resulted in the sizeable conversion of local diluvial uplands and there was significant improvement of riparian engineering, especially in the Kanto Plain. But on the whole, large-scale projects decreased after about 1720, victims of rising costs and natural calamities. Insufficient production and especially inefficient distribution in late Edo times is reflected in the peasant uprisings that mark the early and middle years of the nineteenth century.

### Agriculture after 1868

As might be expected, modernisation, especially industrialisation, had a profound effect on Japanese agriculture, at least in a technical sense. Initially, the gradual disappearance of feudalism introduced fundamental changes in the economic structure. Land ownership and tenancy, and the organisation of rural society were significantly altered, although the effects of these changes are seen more in the economic standing of the farming class than in the appearance of the landscape.

The mechanics of land ownership were revised after 1868, and from then virtually until the period of land reform following the Second World War the farming class of Japan usually occupied a lowly position in the economic hierarchy. Whereas formerly the farmer had at least held a relatively high *social* position, being second only to the *samurai* class, he was now relegated to a socio-economic stratum that was close to the bottom of the scale. Moreover, while the farmer had rarely been affluent in feudal times, he was not a tenant in the usual sense, even though he may not have held title to the land. But after 1868, with the substitution of absolute rights of ownership and a strictly money-based economy, the majority of the rural population became specifically tenants; as such they were subject not only to heavy taxes but also to rents and rates of interest that were even more burdensome than in the past.

Since ancient times intensive paddy cultivation had required dense communities of cultivators who, by tradition, were disciplined to hard work and cooperative labour, and who were organised into closely knit villages made up of large, extended families. The cultivator, also by tradition and by the laborious and exacting nature of his work, tended to retain a strong emotional attachment to the land without owning it outright.[4] Above the cultivator, groups or individuals, either private or governmental, may have held title to the land, but rather than being landlords as such, these proprietors or managers normally received shares of profit, while the political and social organisation of the cultivators was left largely to the cultivators themselves. Since there were no manors in the European tradition, there were no manorial lands around a central residence; nor were there home farms surrounded by fields. Rather, individuals cultivated a number of tiny plots which were typically separated by the topography as well as fragmented into smaller and smaller units by centuries of inheritance based on primogeniture. Holdings were therefore of vastly varying quality. The lack of contiguity and the varying desirability of individual holdings remain to this day serious deterrents to the kind of field consolidation necessary for modern, large-scale mechanised agriculture. Holdings today are not only small and scattered, they are also located at differing elevations, and since there is a continuance of the emphasis on irrigated rice, with its special requirements including fields that are capable of holding standing water, the more valuable plots are usually those corresponding to the original bottom lands.

As a result of these circumstances the typical traditional settlement was a village, with its own corporate identity regardless of the relationships between its members and their individual proprietors, and with its own governmental organisation normally consisting of a council of family heads (whose status was usually dependent on age and lineage), and a headman who, ideally, was the most venerated member of the community. This system, bolstered by such strong traditional values as fidelity, loyalty and frugality, allowed the maintenance of governmental cohesion even though central control was only loosely applied. The mechanics of Edo period rule so improved the efficiency of this kind of organisation that it was used as a base for social control after 1868, thus embodying the principle, often mentioned, that in traditional Japan government control stopped at the village level.

Agricultural developments after 1868 can be divided into those that occurred in the period before major industrialisation, roughly before 1890, and those that occurred thereafter. The earlier period is characterised more or less by a continuation of the growth pattern of the late Tokugawa

## The agrarian landscape

2.3*a* Computer rooms in a large hydroponic farm near Nagoya
  *b* Green vegetables being grown in computer-controlled hydroponic greenhouse

period, namely, the expansion of paddy lands into places that were underdeveloped either because of inaccessibility or because they were in some way marginal. Thus new growing areas in Hokkaido, Tohoku and Kyushu were opened up, and diluvial and other uplands were committed to paddy production in Old Japan. Reclamation projects were resumed with vigour, as these were considered to be good financial risks, and work was furthered by the importation of new machines and techniques from Europe and the United States. Upland conversion to paddy production was fostered by imported industrial equipment and methods, and in particular there was marked expansion because of new means of irrigation. But inevitably other national needs took precedence over the agricultural sector and, particularly after the Sino-Japanese War of 1894–95, large-scale projects were increasingly rare.

Agricultural development in the period of industrialisation after 1890 involved some significant improvements in the product itself, as well as the continued utilisation of new machines, especially for pumping, spraying and so on. Rice strains were much strengthened through experimentation and research, and the spread of paddy cultivation reached places that had been inaccessible because of such factors as climatic extremes or soil deficiencies. Isolated lowlands in such distant and formerly forbidding locations as northcentral Hokkaido thus became successful producing areas during a short summer. The Nayoro Basin of Hokkaido at 44° 10′ north latitude is the most poleward site in the world for the regular production of irrigated rice, and marks the extent of the achievements of the time in this regard. Paddy acreage grew steadily throughout the first three decades of the twentieth century and reached a peak in 1932. The years since have seen some decrease, especially where production areas have suffered competition with other (chiefly urban) forms of land use. Also during these years there was marked development of government-sponsored agricultural experimentation and the dissemination of such information throughout the country.

Generally, however, as the population swelled — and in Japan, as elsewhere, at a faster rate in rural areas — agriculture became less and less able to satisfy the demand. Mass tenancy, endemic debt and attendant ills not only failed to disappear, but were exacerbated by other circumstances. Especially after the First World War, economic conditions in rural Japan grew steadily worse. There was little outward change, but poverty was widespread even before the worldwide depression of 1929. As early as 1927 the short period of relatively liberal government in Japan ended. Civilian rule gave way to increasingly reactionary forces, foremost of which was the army, whose principal source of manpower was rural Japan. In fact, the early move by Japan to the right was prompted heavily by the dissatisfactions of the farmers, who were especially debt-ridden and burdened by problems of poverty, overpopulation, absentee landownership and chronic underemployment.

The shift by Japan after 1931, and particularly after 1937, to a completely wartime footing benefited the major cities economically and helped to improve urban living conditions to some extent, but throughout the 1930s the troubles of the countryside saw little relief. Farmers were often forced to resort to desperate measures, frequently in vain, to try to relieve financial burdens.

## Agricultural growth since 1945

The vigour of Japanese economic expansion in the postwar period, especially after 1955, has been so remarkable, and the burgeoning of the secondary and tertiary industrial sectors so spectacular, that the growth of agriculture and of primary pursuits in general has tended to seem of less significance. Yet agriculture also has realised many unprecedented gains, both in overall efficiency and in the status of its practitioners. The once economically downtrodden farming class, for example, has been almost entirely replaced by a relatively affluent, landowning, rural middle class, and gone are such traditional agrarian scourges as the absentee landlord and the wretched tenant. Signs of affluence in the countryside are everywhere at hand in new buildings, new vehicles and accoutrements, frivolous as well as professional, and in a general aura of wellbeing never before seen in such surroundings. The basic picture may be much the same but many details are new and perhaps hidden, especially to the uninitiated. There is

2.4 Modern farmhouse, Iga Basin, 1977. Design is traditional but materials are often modern. Note solar water heater on new tiled roof

still much validity in a description of small plots, carefully tended and of enormous variety in season, interspersed with tidy but narrow lanes of access, intricate networks of drainage and irrigation ditches, scattered stacks of straw, neatly arranged for later use; of compact settlements of attractive wooden and stucco structures with dark roofs of interlocking tile surrounded or backed by abrupt wooded hills, the whole setting being dominated by distant mountain ranges or by variegated coasts with stretches of island-studded sea. But such a description might entice the reader to ignore such subtle changes as the addition of television aerials, telephone lines, improved roads and irrigation facilities, or such as modernised kitchens with hot and cold running water, gas or electric stoves and other equipment such as water and bath heaters, sophisticated electrical appliances or wholly new electrical systems within the home. More subtle still are the myriad changes that have come to the styles of daily living in farm families. The profession, to be sure, has played a somewhat subordinate role in the general national advancement since the 1950s, for individual incomes have consistently and even increasingly become attributable to urban-industrial employment pursued at first as part-time work but which has now tended toward dominance. But the almost silent revolution within the agricultural domain remains one of the more vivid elements of success in postwar Japan, chiefly because it represents an almost complete reversal of previous circumstances.

The original agrarian reforms, moreover, are clearly related to policies fostered and rushed to completion at the urging of the Allied (or American) Occupation. Without the enormous power and prestige of this organisation in the years 1945 to about 1950, it is highly doubtful that the changes in the lot of the average Japanese agriculturalist would have been as sweeping, or that their effects on the entire history of postwar economic progress could have been so decisive. This was probably the most monumental of all the Occupation reforms, although again it must be said that the changes were more in the economic position of the profession than in the visible restructuring of the landscape.

Actually, conditions for many farmers had begun to improve several years before the end of the war, simply because, as food producers and suppliers of military manpower, rural Japanese were in a better position to enjoy the fruits of their labour. City folk, who at the start of the war had profited by full employment and the amenities of city living, had begun to feel the pinch of a severe rationing system and other heavy restrictions; and as the end of the war approached, particularly in the spring and summer of 1945, it was they who bore the brunt of the devastating bombing raids, chiefly by United States aircraft. Large city populations, which had reached a high point for the time in the early 1940s, were drastically reduced in 1945 and many survivors were forced to flee, usually to rural areas. Almost everyone, it seemed, had family or close friends left on the farm, and in late 1945 and early 1946 urban refugees rushed to the

*The agrarian landscape*

countryside and somehow were squeezed into farm homes. Those who managed to remain in the city, moreover, particularly in the winters of 1946 and 1947, found living conditions inordinately difficult. Food supplies were particularly inadequate during that bleak span of years, not only because of shortages but also because wartime rationing mechanisms had become disrupted; it was usually necessary for family members (notably the women) to make regular foraging expeditions, often by decrepit interurban trains, or by bus, bicycle or even on foot, to nearby rural parts to plead for the privilege of purchasing, with dearly acquired amounts of heavily inflated currency, pitiful quantities of staple and other foods. Some indication of the widespread dispersion of the population that occurred is evident in the figures according to the sizes of settlements, especially between 1944 and 1945. Generally, places of less than 30 000 population are not cities, and it is obvious in Table 2.1 that settlements of this size underwent large increases, while cities of 100 000 or more, suffered almost identical losses.

TABLE 2.1 Percentages of population in cities, towns and villages by size group, 1940—50

| Year | Over 1 million | Size groups in '000s | | | | | | |
|---|---|---|---|---|---|---|---|---|
| | | 500—999 | 100—499 | 50—99 | 40—49 | 30—39 | Under 30 | Totals |
| 1940 | 17.2 | 2.7 | 9.5 | 5.2 | 1.7 | 2.8 | 60.9 | 100 |
| 1944 | 16.3 | 2.6 | 11.3 | 6.5 | 1.8 | 3.3 | 58.2 | 100 |
| 1945 | 5.4 | 2.9 | 7.0 | 7.5 | 2.7 | 3.5 | 71.0 | 100 |
| 1950 | 11.4 | 2.1 | 12.2 | 7.6 | 2.7 | 3.1 | 60.9 | 100 |

Source: *Population Census of 1950*, vol. 1, Table 6, p. 39, with figures rounded.

**Postwar land reform**

As noted above, beginning in the early years of the war and especially in the period immediately following, Japanese farmers had never been so exalted. And while the profits were probably more in their suddenly elevated prestige than in great accumulations of individual wealth, their change of status was documented in legal reforms, promulgated by the Occupation (after some rather feeble attempts by the Japanese in the years preceding) and ratified by the Diet in 1946. The degree of acceptance of these measures is further illustrated by their virtual extension without significant alteration in the Agricultural Land Law of 1952, even though the Occupation had ended and Japan had returned to sovereign nationhood.

The reforms, while sweeping, revolutionary and couched in complex language, were basically simple. The main aims were to eliminate, once and for all, absentee landlordism and such concomitant evils as excessively high rents and rates of interest, and payments in kind for rents and taxes. Hence, with the implementation of the new laws in 1946, the old agrarian diseases which the Japanese farmer had held in common for centuries with his counterparts in rural Asia, were outlawed, and the age-old problem of irreducible debt for the cultivator began to approach solution. The crux of most legislation, including the 1952 Agricultural Land Law, lay in the restriction of agricultural land holdings in Old Japan to no more than 3 hectares (7.35 acres) for self-cultivation and to 1 hectare (2.45 acres) for rental out; and in Hokkaido, since land was generally cheaper and more profuse and the pressures of population less, to 12 and 4 hectares (29.4 and 9.8 acres) respectively. Loopholes in older laws, which had allowed the reversion of some lands to former holders and certain other irregularities, were plugged; and within about two years — mainly between 1947 and 1949 — there had been, firmly but on the whole peacefully, a wholesale transfer of property. The plugging of loopholes which had previously prevented the full effectiveness of such legislation, involved the prohibition of rent payments in kind and, in general, a much stricter control over rents and land transfers. Thenceforth such transactions were cumbersome, requiring the consent of prefectural governors acting on the advice of local land committees elected by members of the communities in question, and it was mandatory to have agreements in writing, since many of the ills of the past had stemmed from loose, oral contracts which always seemed to work to the disadvantage of the tenant cultivator.

The difficulties resulting from the postwar land reform laws were several. Some were soon nullified by rapidly changing conditions, while others persisted longer and were only gradually eliminated; a few have persisted down to the present. In the first category, there was an immediate problem of underemployment in rural areas, for in addition to the usual surpluses of farm labour, there was, between 1945 and 1950, the aforementioned body of urban evacuees which, soon after the surrender, was joined by a sizeable portion of the nearly 6 million Japanese repatriates from other parts of Asia. The ensuing years, however, especially those since 1960, have seen not only the erasure of these conditions but their almost complete reversal, thanks mainly to the ability of other economic sectors to provide job opportunities.

The second major problem involved, and to some extent still involves, the conditions of payment in the late 1940s, whereby properties in excess of the stipulated amounts were purchased by the Government from former landowners. The method of land transfer entailed the sale of these lands at prewar prices, and their subsequent resale at low rates of interest to former tenants. And since between 1947 and 1949 there was a period of major inflation, the former landlords felt that they had not enjoyed the 'just compensation' called for in the law. With much justification they argued that their property had virtually been confiscated. However, attempts to have the law declared unconstitutional were denied in a Supreme Court ruling in 1953, and their demands for some monetary restitution went

*The agrarian landscape*

largely unheeded (despite the increasing vehemence of their appeals) until the early 1960s, when at last the Government agreed, in the face of severe protest by the Socialists, Communists and other anti-Government elements, to allow about $5.5 million in credit for 'deserving landlords' who needed capital for business ventures.[5]

The third problem concerned the very term *landlord*, for regardless of their unsavoury reputation and their undoubted economic power in previous times, the majority were small operators, often individuals who had deserted farming for urban occupations. Widespread abuse *had* been present, but mostly in Tohoku and Hokkaido, where conditions of rural deprivation had been sufficiently onerous to have created a general need for wholesale reform. But on the whole the landlord ranks were composed of thousands of rather ordinary citizens who derived some income from rented farm property and who thenceforth would be denied this. Consequently, the dispossessed element was a large one and its grievances were such as to make it a formidable power bloc, especially to the conservatively-inclined Liberal-Democratic party, which has controlled the Government by a wide majority throughout the post-Occupation period.[6]

The overall effectiveness of the land reform is a moot question. It was recognised as early as the first part of the 1960s that speculation in land for agricultural purposes was no longer very attractive, and that if the law were suddenly repealed there would be little danger of reversion to past circumstances. In fact, the law itself is widely regarded as a barrier to large-scale agricultural modernisation. None the less, changes have come only recently and very grudgingly in the direction of increased ceilings on landholdings, which may eventually cause the modifications in the landscape to which there was some allusion at the beginning of this chapter.

Meanwhile, since the inception of the land reform programme there has been an increase in virtually every phase of agricultural production. Technologically, the reforms coincided with unprecedented developments, not only in fertilisers and pesticides but also in seed selection and methods of cultivation, as well as in the creation of specialised machines and tools. Agricultural cooperatives, whose operations date from the years before the Second World War, were much strengthened in influence and services, and soon there were such peripheral conveniences as local outlets for the sale at discount prices of an expanding variety of products and goods, irrespective of their direct professional connotation. Farm cooperatives also came to be agencies for low interest loans and other financial services, which were added to their main function of providing technical information on the intricacies of raising and marketing a vast assortment of products. Government policy also promoted the agricultural sector, and although special attention was paid (in the form of subsidies and technical assistance) to the rice trade, there was substantial encouragement of crop diversification and of land improvement in general. Consequently, the conversion of diluvial uplands and of such marginal lands as areas of peat bog in Hokkaido were fostered by the new policies, and historic reclama-

2.5 Rice harvesting combine delivered to field by tractor and flatbed trailer. Ise Plain, 1976

tion projects in such places as Kojima Bay of Okayama or the Hachiro Lagoon of Akita, were pursued with renewed vigour, often as adjuncts to the development of new hydroelectric or water control projects.

As a result of these changed conditions the Japanese farmer was given a new sense of energy and purpose as well as a new position of dignity. In the modern period especially, with the growth of urban culture in the Western context, the farmer had tended to be considered a kind of second-class citizen in the eyes of city dwellers or of the educated in general. But the wartime experiences and the effectiveness of the land reform, in addition to the increasing tendency for rural folk to take up urban pursuits, have largely ended such prejudice.

The post-1945 agricultural renaissance is seen to best advantage in the growth of such previously neglected enterprises as cattle and hog raising; in the dairy industry, not only for fluid, canned and dried milk, but for a full assortment of cheeses and related items; and in the production of chickens and eggs by the use of modern, labour-saving techniques which have kept the prices of these commodities so low that these have become attractive substitutes for more traditional dietary fare in terms of the daily food budget. Fruit production, at which the Japanese had long excelled, was immensely developed, resulting in greater quantities, better quality and expanded varieties, albeit at increased prices. Hence grape growing, which

*The agrarian landscape*

in the past had been confined to limited areas (notably to the Kofu Basin of Yamanashi prefecture) has become nearly ubiquitous, and the wine industry, almost non-existent in prewar years, is now well developed and even sophisticated in the kinds and quality of its products.

Rice output rose steadily after the war and reached unprecedented heights until in the early 1960s it began to surpass the demand. Moreover, thanks to increasing amounts of new fertilisers and pesticides, and to new strains and techniques, especially to new kinds of mechanisation, the old cycle of underproduction in certain years because of natural disasters ended after about 1955; each year since then has seen bumper harvests, so that in the fiscal year 1977 the rice harvest surplus was 1.62 million tons, compared to a normal deficit of about 25 per cent in most prewar years.[7]

Of the many changes in the traditional agricultural cycle, probably the most fundamental is the virtual disappearance of off-season (usually winter) crops in favour of imports, mainly from the United States. This development began in the 1960s and is now firmly established. Wheat is the most important of these but it was always usual to see other dry grains such as barley of various kinds, rye or millet being raised without irrigation in the paddy plots following the autumn rice harvest. Now such fields simply lie idle except where demand — usually near clusters of population — calls for the raising of root crops or of vegetables and fruits, often under glass or vinyl. In addition, a number of rather large-scale agro-factories operated by rival firms have been established in various parts of Japan, in which fruits, vegetables, and even such marine products as eels and shrimp are produced by the most advanced techniques including the use of computers and fully-equipped analytical laboratories. In this kind of farming few workers are needed, climatic vicissitudes are entirely avoided, and plant nutrients are automatically circulated in water whose temperature and other properties are carefully monitored by complex instruments.

These unprecedented changes have come about because of increasing emphasis on urban sources of income, a development which has almost wholly erased the feasibility of full-time agricultural activity for the average able-bodied male. Women, the elderly, and the very young might attempt to continue traditional ways but the transition has been general and is probably irreversible. Other reasons for such change lie in the availability and cheapness of foreign imports, and finally in greatly altered public taste, which has called for more and more wheat — for use in bread, noodles, and other common foods — , for soy beans, whose uses are many in both traditional and modern terms, and for fodder crops to support an ever-increasing preference for meat and chicken, as well as for dairy products. Fast-food chains, predominantly American, became well-established in Japan in the early 1970s and since then thousands of competing restaurants and stores have been opened, and all this has helped greatly to augment these trends.

Consequently, except for rice, the level of self-sufficiency in agricultural products has followed a continually downward pattern. And although the latest government policies are designed to reduce dependency on imports, expectations for success within the next decade or so are conservative at best. Japan, it is anticipated, will continue in the foreseeable future to rely on world markets for most products of farm and forest.

Rice harvests, meanwhile, although now scheduled for modest volumes and even — to eliminate chronic surpluses — for some decline, have been maintained by the increasing use of machines in all stages of production. The old seed-bed routine, for example, which saw the young crop produced *en masse* by each farm household in a specially prepared plot prior to being hand-transplanted into other plots which had to be cleared of winter crops and similarly prepared, is now universally transformed. Seedlings are now raised in shallow wooden or plastic boxes which are filled partly by machine — the whole process taking perhaps one morning and occupying the most vigorous male members of households in their spare time. The boxes, which are set out, watered and touched up with more fertilised earth, are then transferred to a kind of incubator and left until the sprouts are well established before finally being placed in a greenhouse where growth is further controlled until the seedlings are about 30 centimetres high and ready for transplanting. Then, each mass of seedlings (a mat of tangled roots and neat, light-green shoots) are loaded on a petrol-powered machine designed for a number of such mats, which usually plants two rows at a time as it rapidly skims (on pontoons) over the flooded plot, the operator walking behind. And since men customarily operate machines, women assist by setting up boxes and making other preparations. At harvest time, machines have also eliminated most of the 'stoop labour' and have allowed a quicker, more efficient completion of the ancient cycle. Investment in machinery, on the other hand, has been unusually brisk, especially as the sharing of major implements within villages has virtually ceased, each household now doing its best to provide its own equipment in order to maintain independence in the cycle of production.

## Some contemporary problems of agriculture

In spite of the many improvements in the agricultural sector over the past thirty years or so, there have remained or have arisen a number of problems, only some of which are related to prewar circumstances. In fact, most of Japan's agrarian difficulties, as with many of its other contemporary problems, are entirely like those of the industrially advanced nations in general.

The death of cultivable land, the intrinsically poor quality of the soils and the vagaries of the climate are, of course, continuing problems with long histories. And the still unsolved problem of small field size, rendered

## The agrarian landscape

smaller through the ages by traditional inheritance practices, is a major barrier to large-scale mechanisation and a universal shift to extensive methods. Cultivable acreage, moreover, has suffered further reduction in the postwar era because of urban encroachment, but poor soils have been corrected to some extent by the liberal application of more effective fertilisers and other ingredients — albeit since many of these are derived from petroleum, at higher and higher cost — as well as by the use of new equipment for field preparation and maintenance, while the climatic variations, as suggested, have been more effectively controlled by new techniques and materials or by stronger strains developed through agricultural research.

The problem of inadequate return for agricultural activity also smacks of the past but is no longer attended by such traditional consequences as rural poverty, thanks to widespread opportunities for supplemental income. Indeed, in 1977 only 648 000 farm households (about one-third of the total) were recorded as being *exclusively* engaged in agriculture.[8] As this figure indicates, a decreasing supply of labour, particularly of young males, is a problem that is not only recent, but is one that has wide and permanent implications in that the profession is no longer attractive, at least to a large segment of the public. As there is increasing need for agricultural mechanisation on a broad scale, wholesale change in the agrarian landscape seems inevitable. Some, such as the disappearance of winter field crops, has already occurred.

The greatest changes at present being made in the agrarian landscape are those in conjunction with expanding urbanisation. Large areas round most cities are rapidly (and often in disregard of zoning regulations) being converted into industrial sites or housing projects, as noted in Chapter 3, and although such growth has occurred generally on reclaimed lands or in hilly areas, there has also been absorption of good paddy land for these purposes. The sale of property to speculators has been a special boon to many farmers, particularly to those bent on a secure retirement, and it is no wonder that the growing number of Japanese tourists travelling abroad includes an ample proportion of elderly men and, especially, women, whose backgrounds are entirely rural.

Much obvious change in the rural landscape has taken place since the early 1970s because of the enormous growth in popularity of the family car. Main roads, even in hitherto completely rural settings, have thus become heavily committed to automobile service agencies of all sorts, including such ancillary facilities as supermarkets, drive-in restaurants, *pachinko* (pin-ball) parlours, and motels. The highway traveller is thus often obliged to view the scenic beauty, for which the culture has always been famous, in segments, as these may still be seen between clusters of highly diverse and often blatant features which aim to attract the passing motorist.

2.6 Small family harvesting rice by combine

## Summary

The Japanese agrarian landscape still retains an air of traditional, meticulous beauty, but changes are occurring rapidly and the future scene may well seem more familiar to Europeans and Americans than to other Asians.

The history of agriculture in Japan pre-dates written record, the first descriptions of the culture revealing a well-established agrarian base for the sophisticated and largely urban court life. The emphasis from earliest times on irrigated rice was greatly accommodated by the kinds of marshy, alluvial lowlands found in Japan which, although limited in extent and number, were initially compatible with the culture that evolved. The rice cycle as the foundation of all agriculture developed as a consequence of a generally mild, moist climate, of which the alternation of dry and wet periods and of chill winters and very warm summers, has comfortably encouraged the stages of production.

Gradually, as the Japanese farmer learned to deal with the adverse aspects of his environment, the agricultural domain was extended to include marginal hill and submerged lands, much of which was levelled, diked and outfitted for paddy culture. Lands unfit for such reclamation were generally farmed much less intensively, usually to supplement the growth of the primary staple. All lands came to be used throughout the year, however, for various basic crops, with multiple and intercropping in season, providing a variety of items under conditions of increasingly close land management. Human excrement, animal and vegetable fertilisers were early used to enrich the intrinsically poor soils, and yields were generally high.

Each historical period saw agricultural advances, and these were favoured by imported tools, techniques and expertise. The Koreans especially provided much early assistance and influences from China, although perhaps less immediate, were equally important.

The first 120 years of the Edo period saw particularly strong developments in agriculture. New lands were created until virtually all space available for rice had been brought into production. The economic foundations of the state, and of the some 260 fiefs, lay in their capabilities for rice production, and for most of this time rice was the medium of exchange. Since productivity of feudal lands formed the bases for the political rankings of the *daimyō*, there was strong competition and impetus for improvement. After about 1720 activity slowed and there were sporadic crop failures which, especially in the early nineteenth century, led to local unrest.

The Meiji period brought much technical change to Old Japan and inaugurated large-scale rice production and subsidiary agriculture in Hokkaido, but generally modern times, with a strict money economy, inflicted hardship on the farmer. Land came to be owned by a relatively few and often by absentee landlords, while the majority of the population was reduced to the status of debt-ridden tenant operators. The Taisho (1912–25) and early Shōwa (1926–39) periods saw little improvement, but the farmers benefited in some measure by the war, although the basic ills of tenancy and debt were not ended until after Japan's defeat in 1945.

The years 1946–49 were a period of revolutionary land reform which reversed the previous conditions and eliminated tenancy. Meanwhile, after 1945 the exodus of city folk and their resettlement on farms helped to create a new feeling for the agriculturalist and to give him unprecedented status, both economically and socially. Since then, in terms of personal income, farmers have been outcompeted by urban occupations, but most have drawn heavily on employment opportunities in the city to augment and finally to dominate their farm incomes. 'Part-time' farming, performed by commuters, is the rule in Japan today, with routine chores being accomplished by women, children and the elderly. But since 1945 there has been a direct reduction in the farm population. Agriculture no longer attracts the young, brides for farm families are almost unobtainable, and the sale of farm property for urban uses has become a prevailing trend. Yet, by virtue of mechanisation and technology, rice production has continued to soar and surpluses create new and hitherto unknown problems. Other traditional commodities are supplemented more and more by imports, but there has been a marked shift to new products in answer to changing tastes and general affluence, and the government is making plans to encourage the production of crops other than rice.

*Summary*

Although postwar land reform legislation still controls property ownership, a relaxation of existing laws had been noted in recent years. Ultimately there should be an increase in average holdings to allow for field consolidation under single management and the use of truly large-scale techniques. In all probability, corporations will eventually replace the family in the general business of agriculture.

# 3
# The city in Japanese history

While Chapter 1 explored to some extent the general topic of the city in Japan in terms of the physical setting, and of the growth of a city-oriented culture into the predominating way of life in the present day, the forces which were considered to have brought this about, and even in some measure those that are purely environmental, are essentially attributes of the modern period of history. Modern in this sense refers to the time since the so-called Meiji Restoration of 1868, when the control of government was wrested from the hands of factions opposed to the peaceful transformation of Japan into an avid student of the nations of northwestern Europe and North America. These forces are thus chiefly economic, deriving from Japan's ambition to become a world industrial and commercial power, and from its remarkable success in this undertaking.

On the other hand, the development and growth of cities in Japan, and the establishment of an urban hierarchy that in many ways has persisted to the present, were not simply the results of post-Meiji economic and political forces, coupled with appropriate adaptation to the peculiarities of the physical surroundings. The groundwork for the modern urban system was, in part deliberately, begun at least 250 years before the Meiji Restoration, and the concept of the city was recognised and had been effectively subjected to the inevitable process of acculturation as early as the Heian period (AD 794–1185), and perhaps even before.

The role of history, therefore, has been a factor of great importance in the entire evolution of a city-oriented culture in Japan, and while obviously only the bare outlines can be presented here, it is vital to the context of this resumé to touch upon the highlights of this development. In any case, the reader can be reminded that the city, as such, in Japanese culture, was widely known to a large portion of the public long before the establishment of a Japanese nation-state, and hence remotely before the entry of Japan into the modern world of international trade and commerce.

## The beginnings: AD 710 to 1572

Recorded history for the Japanese begins after the mid-sixth century AD, when Buddhism was introduced from China and Korea, and its intellectual and other attributes, including written Chinese, had been absorbed. Before

## The beginnings: AD710 to 1572

this, from the written records of others (since the Japanese had not developed a writing system) and from archaeological evidence, it is known that the archipelago was peopled by various elements, notably a so-called proto-Caucasoid strain (the Ainu), and their principal enemies, the main ancestors of the modern Japanese. There were others, such as the Koreans and Chinese, who had become sporadically admixed, but by the time of the introduction of Buddhism the early Japanese had come to dominate much of Old Japan, at least from the area around Lake Biwa westward, with the Ainu resisting stubbornly in remote places, particularly in the northeast. The conquering of Honshu was a slow process that occupied the Japanese for many succeeding centuries, while their own culture was often being strained by internal rivalries and, late in the thirteenth century, by two massive but vain assaults by the Mongols. The great island of Hokkaido to the north, meanwhile, except for its most southern extremities, remained outside the fold until after the Meiji Restoration of 1868.

The strongest of the ancestral Japanese chose the fragmented lowlands between Osaka Bay and Lake Biwa to be the seat of their culture, which was clan-oriented and dominated by a militaristic element whose strength and ascendancy, in part, were achieved by superior strategy, mobility, and the use of such advanced weapons as the long-bow and superb quality steel swords. Beneath the ruling military class the basic and stable element was agrarian, with paddy culture as its central activity.

Similar people with similar ways existed in western Japan, especially in the fertile lowlands of riverine deposition that lay along the shores of the Setonaikai and in northwestern Kyushu, but by the sixth century the lowlands around what is now Osaka Bay, eastward to Lake Biwa and including the Kyoto and Nara Basins, seemed the most appropriate for the expansion of the young culture.

The implantation of Buddhism acquainted the Japanese not only with a religious system that was perhaps more sophisticated than their own animistic *Shinto* belief, but also, along with a flood of Chinese influences, came a new way of life that included elaborate social and political organisation, art, architecture, music and literature, land reform and city planning, and even a completely new ethical order, Confucianism. All these were gradually superimposed on the functioning culture which then obtained. Eventually, although at first the Japanese may have adopted Chinese forms (which were transmitted to them by Chinese and especially by Koreans, as well as by their own emissaries) without much change, most elements ultimately became Japanised and greatly altered.

Before the coming of Buddhism, settlements in the areas subdued by the Japanese had been mostly transitory. From prehistoric times it had been the practice to move to new headquarters with the deaths of clan chieftains, who then became enshrined in large tomb mounds (*kofun*) that are often recognisable landscape features. No settlements worthy of the name *city* remain from the time before the completion of Heijokyo (the ancient city of Nara) in AD 710, although some elaborately planned provincial towns, such as Dazaifu in northwestern Kyushu, are known to have flourished. It may indeed be an exaggeration to consider the modern city of Nara an example of early Japanese city building, for today it is but a relatively small and rather insignificant prefectural capital which reflects the grandeur of the Nara period (AD 710–84) only in the magnificent shrines and temples which so impressively punctuate the beautifully landscaped hills at the eastern edge of the city, and which form one of the most prominent tourist attractions in the nation, for Japanese and foreigners alike. The splendour of the city itself has virtually disappeared, but the plan is known, and the place of the city in the culture is well established.

On the other hand, while Nara was executed by the Japanese, its plan was actually Chinese in conception. By the time of its realisation, Buddhism had become firmly inculcated, and the force of the T'ang civilisation had flowed freely into Japan, colouring all aspects of life, especially among the privileged. A counter trend, characteristic of the Heian period which followed, had not yet taken firm root, and the Japanese were still in the flush of an inspiration which was basically Chinese, but which gave them the background for masterful and uniquely Japanese creations in succeeding centuries, and earned for them a richly deserved reputation for artistic expression and good taste.

Much the same might be said of the basic plan for Heiankyo (modern Kyoto), although, unlike Nara, the original outlines of the city are still evident and Kyoto is therefore a true living monument, even aside from its religious structures. But such early city plans were, in the eyes of the writer, fundamentally alien to the Japanese temperament, both artistically and physically, in that they had been based on plans for the T'ang capital of Ch'ang An and were therefore symmetrical in a way that is not usually associated with the Japanese. The basic pattern is one of a rectangular grid with a series of broad avenues arranged at regular intervals around a palace in the upper centre, the whole placement being rigidly fixed according to the cardinal points of the compass. It seems obvious that the very nature of such a design is out of keeping with the seeming artistic proclivity of the Japanese for asymmetrical forms, and thus the whole expression appears to represent something that is primarily non-Japanese. Nor is such a plan physically suited to a country like Japan, with its irregular surface configuration and dearth of level land. Even in modern Kyoto, where the Kyoto Basin has provided a relatively ample expanse of flat land, it is quite evident that the physical limitations of the site have precluded the perfect execution of the initial design.

Therefore, although Kyoto and part of Nara are of exceptional interest historically, their influence on the morphology of the Japanese city as it developed in later centuries is rather negligible, and if they can be said to have left their mark in modern times, perhaps it can be seen in the pattern of city streets in the central business areas of modern cities, which are rather uniformly rectilinear.

## The beginnings: AD 710 to 1572

The Nara and Heian periods did produce, of course, a most sophisticated civilisation, as can be seen for example in the English translations of the contemporary classic, *The Tale of Genji*, or in the modern historical novel, *The Heike Story* by Yoshikawa Eiji. Such a society, although it was elitist and thus significant of a relatively small portion of the population, was, in its sociopolitical organisation and even in its economic structure, thoroughly divorced from the primary industrial sector. These early periods also embodied an administrative system, compounded and ramified from original Chinese models and codified in Japan in the period of the so-called Taika Reforms (AD 645–701), which required the formation of centres of communication. Hence, though it took many centuries for the successful realisation of political control on a national scale, there gradually emerged a network of permanent nucleated settlements under a central authority, and the experiences gained in the times of early cultural flowering under the impact of Buddhism made valuable contributions.

After the Heian period at least four centuries elapsed during which the pattern of city growth was marked by the rise and fall of strategic centres which reflected the temporary dominance of one warring clan or another. As the fortunes of war fluctuated, so did the importance of these centres. None remained powerful or influential for very long except Kyoto, the permanent capital or imperial seat from AD 794 until 1868, and even Kyoto was devastated frequently in struggles involving rival clans or militant Buddhist factions. Some of the *de facto*, or military, capitals, were impressive in their day. Kamakura, for example, although it is now functioning as an important Tokyo or Kanto area satellite city, still exhibits relics of its days as the seat of the Minamoto and Hojo regencies (1185–1333). But not many have remained to become the nuclei of important modern cities. For one thing, their sites were usually chosen for strategic reasons and were therefore deliberately made inaccessible. Also their influence, owing to the strength of their rivals, was never universally felt. In general these were usually merely transitory strongholds, and though some have survived as tourist centres, minor ports, market towns, shrine sites, satellite cities with modern functions, or, at best, as prefectural capitals, the net effect of this pattern of development on later forms of the city was relatively slight.

From the end of the Heian period in the late twelfth century, until 1868, the sociopolitical organisation of Japan is generally known as feudal, with increasing centralisation of power, especially after about 1570. Possibly as a side effect of this condition, there arose a class of fortified and semi-autonomous towns that flourished on overseas trade and piracy, and which have been likened to the free cities of medieval Europe. Foremost among these was the port town of Sakai, a few kilometres to the south of the present site of Osaka, and from early times accessible to deepwater shipping. Osaka, though better situated with regard to the principal trade routes between Kyoto and the west, was long denied much foreign commerce because of a shallow harbour — a shortcoming that has only

3.1 Canalised distributary of the Yodo River at Osaka today: such waterways are still heavily used by barges

been eliminated in modern times by the rigorous application of the most up-to-date techniques of dredging and harbour construction. While at first Osaka harbour was adequate for the shallow draught vessels of the day, the development in Heian times of farming along the Yodo River, largest of the local streams and the main route of Heian travel to and from Kyoto, so accelerated the silting that the use of out-ports became an early necessity.[1]

These natural conditions, coupled with the expanding China trade and, consequently, the increased size of ships, greatly stimulated the growth of Sakai and other deepwater ports of the area. It was Sakai, however, partly because of its strategic position and peculiarly favourable political circumstances, that prospered the most between the fourteenth century and the seventeenth. It became the leading Japanese port of the time, and eventually grew so powerful that, in the words of George Sansom, 'by 1543 we find the *Bakufu* [feudal Government] borrowing money from Sakai merchants on the security of taxes from Ashikaga [historical period 1336–1568] domains'.[2]

Because of such circumstances, Sakai had an unusual degree of local autonomy, including its own militia for protection against invasion during the constant internecine warfare that marks the years of its prosperity. The city itself was composed of dwellings of indifferent design, and of many functional structures, such as warehouses and rudimentary factories. And, since the anti-foreign edicts were not promulgated until about 1640. the population included foreigners, both European and Asian.

None of these ancient 'free' cities has retained its original function or importance, although Sakai is now a bustling and major suburb of Osaka, and Hakata, the second of the class in prominence, is now part of the great city of Fukuoka in northern Kyushu. The demise of this type of city was largely a product of the policies of the three great military leaders of the sixteenth century, Oda Nobunaga (in power, 1568–82), Toyotomi Hideyoshi (Regent, 1582–98) and Tokugawa Ieyasu (*shōgun*, or supreme commander, and the first of the family that ruled Japan from 1600 until 1868). In the eyes of the military leaders, who were gradually forging political unity in Old Japan, the independence of such cities alone was sufficient to warrant their destruction; thus the 'free' cities as a class can be added to the list of pre-Tokugawa settlements that failed to have lasting influence on the urban pattern of the present day.

Meantime, while most of the pre-Tokugawa cities were living out their importance, Kyoto remained a place of singular greatness, although in its some 810 years as Japan's largest and most imposing city, there were frequent periods of decline.[3] It seems to have been particularly ravished during the Onin War (1467–77), yet by the middle of the sixteenth century visiting Portuguese traders reported it a city of 96 000 houses,

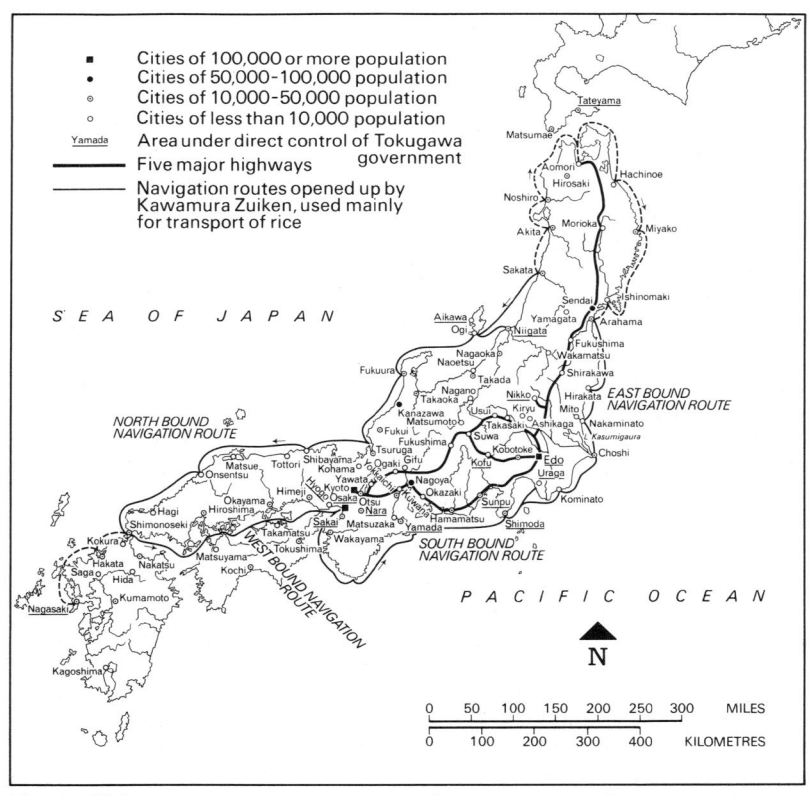

Map 12 Major cities and travel routes in Japan during the Edo period. (*Source*: Yazaki, 1968)

with a population of over half a million, making it larger than any European city of the day.[4]

## The establishment of a permanent network of cities: 1573 to 1868

When the Tokugawa *shōgun*, principally Hidetada (1605–22) and Iemitsu (1623–51), had succeeded in consolidating the gains of their powerful predecessors, a long era of armed peace and comparative but sporadic prosperity began, and it was during this period (Edo, or Tokugawa, 1600–1868) that the foundations of the modern city network became permanent. On these foundations, once the doors were opened to the material advances of the West, the Japanese rapidly built a modern system of cities. That they were able to do this in less than a century after 1868, although certainly a remarkable achievement, is somewhat less so when the origins are examined, even roughly.

While these developments did not come about perhaps as the outgrowth of a brilliant and farsighted policy beginning about the time of Oda Nobunaga, neither did they evolve by accident. A series of conditions, some military or strategic and some political or economic, seem to have made this course inevitable. The gradual unification of the military power which by 1615 had led to a degree of political stability under the Tokugawa *shōgun*, brought also a long and unprecedented period of dynamic growth and change. The cities of the land were to be particularly affected by a new and different sort of urban mechanism, with the castle town (*jōkamachi*) as the symbol of the regime's military and political authority. Edo soon became the chief castle town, and its extraordinarily rapid expansion after being formally made the political capital in 1603 was deliberately accomplished by the regime as a mark of its strength and power.

The castle town was not solely an innovation of the Tokugawa regime; the tendency for a town to develop around a military garrison had become fairly well established, especially during the sixteenth century. There was an important difference, however, in the sort of town that developed after this time, for until the late sixteenth century, at least, the basic function of the castle communities had been strategic. But gradually, as the towns expanded or, more important, as local warfare ceased to provide a need for such considerations, economic and administrative necessities came to exercise a stronger influence on development. Perhaps this can be seen, for example, in Hideyoshi's recognition of the immense potential of the site of Osaka, not only as a military, but also as a great economic and administrative centre. Hence in 1583 he built a grand castle there, and around it gathered, in addition to his military retinue, strong financial interests from such places as the neighbouring and prosperous city of Sakai. The weakening of rival cities, grown too independent in the eyes of the political leadership, was common practice, and this was accomplished in various and mostly indirect ways, such as by luring away their financial sup-

3.2 Osaka Castle, a reconstruction of what was once one of the grandest and strongest castles in Japan. The original castle was built by Toyotomi Hideyoshi in 1583

porters, by undermining their very existence through basic policy changes, or even by rearranging site characteristics (as at Sakai) by such means as changing the directions of river courses, so as to inundate vital areas. In the case of Sakai also, and other of the 'free' cities, their enfeeblement by administrative proclamation is seen in the Tokugawa policy of seclusion, which, in prohibiting most foreign trade after the mid-1630s, virtually assured their stagnation and decline.

Tokugawa Ieyasu and his descendants, thus following a pattern established by their forebears, especially by Hideyoshi, sought to insure the primacy of their rule through a series of edicts that, incidentally perhaps, greatly furthered the growth of cities. One of these regulations, promulgated in 1612, ordered the destruction of minor feudal castles, and since this resulted inevitably in the migration of civilians also to the more important castle communities, there naturally occurred a consolidation of settlements. Another such administrative order, closely following, was a law which prohibited the construction of more than one castle to a province, and which, of course, had far-reaching effects in bolstering the importance of the nodal communities.

The carefully guarded tranquillity of the Tokugawa period quickly brought a change in the standards for castle locations. Of the structures erected, repaired or altered between about 1575 and 1616, for example, many came to occupy sites that formerly would have been considered inappropriate, and that these sites were congenial to the development of modern cities is indicated by their subsequent evolution. As can be seen in Table 3.1, half the cities that grew around these castles (thirty of sixty) are modern prefectural capitals, the great majority are places of some importance as population centres, and most are cities of long standing.

Although the stability brought about by the Tokugawa Government, especially in the choice of castle sites, permitted a downgrading in the need for defensive security in favour of such considerations as the improvement of transportation and trade, the castles themselves were stronger fortresses than before. The buildings, though on a plain, were usually perched on a small hill or locally prominent height, or were elevated to command a view of the surroundings. Castle compounds were protected by moats and by stone masonry partly inspired by Portuguese influence, and also, of course, by the buildings of the town itself.

Geographically, the new castle sites were probably the most ideal in their respective regions for future city growth, as they were generally located on level land and oriented round such landscape features as river-crossings, protected harbours, mountain passes or other prominent access points, or, wherever possible, where there was a combination of several of these features. Equally important, the selection of a site was strongly influenced by the productive potential of the surrounding area, a consideration that would virtually guarantee a certain prominence to the location. The prosperity of the castle town was thus subject to the ability of its hinterland to produce rice and other commodities, and its main function was therefore administrative and commercial, rather than strictly military. For these reasons, the towns were located on flat land and were made accessible to other places, especially Edo, by a complex highway system.

The local administration of the governmental system of the Tokugawa *shogun* was largely in the hands of the *daimyō*, of whom there were around 260, with about a quarter commanding important fiefs (*han*). It is calculated that at the end of the Tokugawa reign there were 258 fiefs, of which fifty-six had rice allotments (the prevailing form of income from the central Government) of over 100 000 *koku* (174 888 hectolitres) per year.[5] This would make them the most important, for since the accounting of rice allotments was carefully recorded (rather than the population and other vital statistics), it is also possible to ascertain by this means the ranks of the various fiefs in the feudal hierarchy. The statuses of *daimyō*, and consequently their incomes, generally depended on such considerations as family ties, and in turn, on past loyalties to the *shogun*, and hence the sizes of fiefs, or more importantly their respective rice-producing capacities, heavily depended on this relative political importance. Thus, the growth of Tokugawa cities and their ultimate sizes and ranking in feudal Japan (and often into the modern age) were largely governed by this 'priority' system which, nevertheless, usually has solid geographical foundations.

One of the more interesting Tokugawa administrative arrangements, and one that had a particularly strong bearing on the rise and growth of cities, was the so-called system of alternate attendance (*sankinkotai*), which required each local lord to spend part of his life in Edo. The expense of periodic and lavish expeditions to and from the provinces, added to the cost of maintaining two residences (families were made to remain in Edo), prevented uprisings by keeping the *shogun*'s potential rivals in a constant state of financial impotence, the ultimate effect being to foster the rapid rise of Edo, and to strengthen the foundations of the modern network of cities. Regional specialisation in the manufacturing of the times, or in the production or processing of agricultural items, although of long standing, was much enhanced, along with commerce, trade and transportation generally. Hence, while Edo became the nerve-centre, Osaka (succeeding Sakai) became noted for business, finance and manufacturing; Nagoya for cotton textiles and, thanks to its favourable location between Edo and Osaka, for commerce as well, and so on, in somewhat the same order as in the modern period. Kyoto, meanwhile, though still the imperial seat, became a kind of cultural shrine city, although it too produced traditional silks, pottery and the like, for which it has remained famous.

The effects of the *sankinkotai* system in promoting the development of cities through a vigorous flow of traffic, however, may not have been as noteworthy as might be anticipated, for the entire process was held in partial check by feudal restrictions. The regime was exceedingly cautious, for example, about allowing a free flow of goods and services, and although highway development and maintenance were notable characteristics of the times, even to the extent of drawing praise from the few foreign visitors, who compared Japanese roads most favourably with those of Europe, the kinds of traffic in general were closely regulated by having to pass through numerous toll and surveillance gates along all highways. These and other difficulties of travel are exemplified by conditions along the great Tokaido route from Edo to Kyoto where, in addition to the toll

TABLE 3.1  Castles, built or repaired, 1575–1616, and present city status

| Date | City | Chartered | Prefecture | 1978 Population ('000s) | Date | City | Chartered | Prefecture | 1978 Population ('000s) |
|---|---|---|---|---|---|---|---|---|---|
| 1575 | TSU | 1889+ | Mie | 140 | 1589 | HIROSHIMA | + | Hiroshima | 853 |
| 1576 | Kuwana | 1937 | Mie | 83 | 1590 | Edo (TOKYO) | + | Tokyo | 8 266 |
| 1580 | Mihara | 1936 | Hiroshima | 86 | 1591 | KAGOSHIMA | + | Kagoshima | 482 |
| 1580 | Himeji | + | Hyogo | 440 | 1591 | Okazaki | 1916 | Aichi | 245 |
| 1581 | YAMAGATA | + | Yamagata | 226 | 1592 | Ueno | 1941 | Mie | 60 |
| 1582 | Matsumoto | 1907 | Nagano | 187 | 1592 | MORIOKA | + | Iwate | 223 |
| 1582 | Kishiwada | 1922 | Osaka | 176 | 1592 | Wakamatsu (Aizu) | 1899 | Fukushima | 111 |
| 1583 | OSAKA | + | Osaka | 2 625 | 1593 | MITO | + | Ibaraki | 205 |
| 1583 | KANAZAWA | + | Ishikawa | 397 | 1594 | OKAYAMA | + | Okayama | 531 |
| 1585 | WAKAYAMA | + | Wakayama | 397 | 1595 | KOFU | + | Yamanashi | 198 |
| 1586 | TOKUSHIMA | + | Tokushima | 244 | 1595 | Ogaki | 1918 | Gifu | 141 |
| 1587 | Kurume | + | Fukuoka | 209 | 1596 | OITA | 1911 | Oita | 333 |
| 1587 | KOCHI | + | Kochi | 289 | 1597 | Marugame | + | Kagawa | 69 |
| 1587 | MAEBASHI | 1892 | Gumma | 258 | 1598 | Takasaki | 1900 | Gumma | 218 |
| 1587 | Shibata | 1947 | Niigata | 76 | 1600 | SENDAI | + | Miyagi | 617 |
| 1587 | Yanagawa | 1951 | Fukuoka | 46 | 1600 | Ueda | 1919 | Nagano | 109 |
| 1588 | Matsusaka | 1933 | Mie | 112 | 1600 | MATSUE | + | Shimane | 129 |
| 1588 | TAKAMATSU | 1890 | Kagawa | 305 | 1600 | FUKUOKA | + | Fukuoka | 999 |

TABLE 3.1 — continued

| Date | City | Chartered | Prefecture | 1978 Population ('000s) | Date | City | Chartered | Prefecture | 1978 Population ('000s) |
|---|---|---|---|---|---|---|---|---|---|
| 1601 | FUKUI | + | Fukui | 233 | 1604 | Hagi | 1932 | Yamaguchi | 53 |
| 1601 | KUMAMOTO | + | Kumamoto | 491 | 1607 | FUKUSHIMA | 1907 | Fukushima | 254 |
| 1601 | TOTTORI | + | Tottori | 125 | 1608 | SAGA | + | Saga | 160 |
| 1601 | Obama | 1951 | Fukui | 34 | 1608 | Yonezawa | + | Yamagata | 92 |
| 1601 | Tsuruoka | 1924 | Yamagata | 99 | 1609 | Takaoka | + | Toyama | 174 |
| 1601 | UTSUNOMIYA | 1896 | Tochigi | 359 | 1609 | SHIZUOKA | + | Shizuoka | 453 |
| 1602 | Kokura* | 1900 | Fukuoka | 386 | 1610 | NAGOYA | + | Aichi | 2 078 |
| 1602 | Sano | 1943 | Tochigi | 78 | 1610 | Hirosaki | + | Aomori | 171 |
| 1602 | AKITA | + | Akita | 273 | 1614 | Odawara | 1940 | Kanagawa | 175 |
| 1603 | MATSUYAMA | + | Ehime | 387 | 1614 | Uwajima | 1921 | Ehime | 72 |
| 1604 | Hikone | 1937 | Shiga | 88 | 1614 | Takada‡ | 1911 | Niigata | 125 |
| 1604 | Tsuyama | 1929 | Okayama | 80 | 1615 | Taira† | 1937 | Fukushima | 341 |

CAPITALS = Prefectural capitals

+ = First charter year (1889)

* = 1978 population (in thousands), of Kokura-ku, Kita-Kyushu Shi

† = Incorporated into the city of Iwaki, October 1966

‡ = Incorporated into the city of Joetsu, April 1971

*Sources:* Toyoda Takeshi, *Nihon no Hōken Toshi* (feudal cities of Japan), Tokyo, 1954, pp. 528–33. Chartered dates and 1978 populations from *Zenkoku Shi, Chō, Son Yōran* (National City, Town and Village Yearbook 53), Tokyo, 1978.

barriers, rivers were deliberately left unbridged, forcing travellers to undergo time-consuming and often dangerous conditions in fording the many streams that cut across the highway. Since these are all the lower courses of typical Japanese rivers, they have been (in the geomorphological timetable) fashioned into old age rather rapidly, thanks to their generally short courses, with headwaters in nearby mountains where the configuration is one of youth. In their short dash to the sea, the streams, always shallow and rocky, and periodically swift and treacherous, plunge within brief distances from mountain source areas across narrow diluvial (elevated, ancient alluvium) terraces, on to newer alluvial plains. Thus the lower courses have been vastly broadened by recurring shifts in stream beds, which have also been elevated by eons of deposition. These *ceiling* rivers are, in addition, contained by pronounced natural levées that, through time, have been heavily augmented by man. Finally, as the streams enter the sea, there is usually great extenuation in the formation of deltas. The negotiation of these features, which along the Tokaido are met in their most extreme form, was, in season at least, particularly hazardous.

Except for special dispatch riders of the regime, travel speeds were extremely slow. Important people were usually carried in palanquins or on the backs of horses, or even of humans (the latter being the chief means for fording streams), but mostly travel was on foot. Hence, even trips of relatively modest distance were tedious and costly.

However, notwithstanding these and other inconveniences, the enforced peace of the times and the compulsion of the military class to travel in large numbers to and from Edo, resulted in a significant surge of activity along highways, especially at nodal points. Consequently, city growth was fostered (as it had to be as an integral part of the regime's administrative mechanism); but again, feudal regulations attempted to control the process. Settlements grew rapidly in early Edo times and populations were often surprisingly large, although there was much fluctuation with the fortunes of a place and particularly with the changes in the political status of the local lord *vis-a-vis* Edo. Migration from the countryside to the growing communities, despite the feudal restrictions, was a leading cause, especially by other than first sons, since primogeniture was the custom and only eldest sons were assured of inheriting the family property.[6]

Population counts on a relatively systematic basis were begun as early as 1721, but for various reasons, including a lack of standard procedures, coupled with security measures, which, for example, did not allow the counting of the military class (the *samurai*), whole segments were often omitted and the results are significantly inexact. It is generally known by extrapolation, however, that the total population remained comparatively stable from 1721 to about 1850, at around 25 to 27 million, but that it grew rapidly before this from a base of about 18 million in the years 1573–92, to 26 million in the period 1716–36, although undoubtedly it was somewhat higher in each case. Of the total, it is estimated that 10 to 15 per cent, an unusually high figure for that time, represented an urban

## The establishment of a permanent network of cities: 1573 to 1868

segment, composed chiefly of those in castle towns.[7]

Edo, of course, was the largest city, and its population, though the count is again imperfect, can be reasonably estimated by combining historical figures with the more precisely recorded rice allotments (*kokudaka*), a standard technique for estimating premodern populations in Japan. By these means Edo is calculated to have had a population of well over 1 million in the latter part of the eighteenth century, and this would have made it the largest city in the world of that day.

More significant still from the standpoint of the theme of this book is the realisation that the urban life of feudal Japan was by no means a one- or two-city affair. Indeed, by the eighteenth century a complex arrangement of settlements had spread throughout the realm. This encompassed literally dozens of towns whose populations are estimated in the tens of thousands, with a goodly number having upwards of 50 000 and the leading few well over 100 000. Next to Edo the biggest were Kyoto and Osaka, both having from 300 000 to 500 000, and these were followed by Nagoya and Kanazawa in the 100 000 class, with such far-flung cities as Kagoshima, Hiroshima, Okayama, Sendai, Nagasaki and others at from 40 000 to 70 000, and so on down the hierarchical scale.

Premodern Japanese cities can also be classified according to function, and although as has been stated, the castle town was supremely important throughout the Edo period as the very hub around which revolved the vital administrative machinery of the regime, there were several other types of cities which, regardless of their origins as castle towns, functioned somewhat differently.

Osaka, for example, along with such prominent cities as Kyoto, Shizuoka, Nagasaki, Kofu, Sakai and others, was kept under the direct control of the Edo Government. This was for security reasons, as these places were known as centres for espionage, but the distinction was also functional. Usually, such places occupied a key position in the intricate arrangement of towns and lines of intercommunication by land and sea for the production, collection, storage and distribution of commodities, foremost of which was rice. This was first sent to Edo for redistribution to the domains, but it followed a circuitous course from the main source areas in northwestern Honshu (the Ou district). Osaka was the most famous commercial centre, and residence there was restricted by orders from Edo to persons of that calling. Those of military rank were excluded, and since this class was regarded as the elite element of society, Osaka culture came to be associated strongly with business, as opposed to the more aristocratic (and, in the popular mind, perhaps somewhat more 'pure') civilisation of Edo. Even in the present day, despite the temporal nearness of the two giant cities, the stigma of an Osaka speech form and its associations, as contrasted to the more favoured Tokyo or 'standard' dialect, persists. Educated parents all over the country and even of Osaka itself are careful to train their children in the use of the Tokyo vernacular. And if a person has, by birth and residence, descended from three generations in Tokyo, always a rarity but even more so today with the accelerated pattern of immigration, he is liable, if asked, to announce with no little pride that he is an *Edokko*, or a genuine person of Edo. Cultural regionalism in Japan today, although weakening, is still relatively widespread, and the dichotomy between Tokyo and Osaka is probably its strongest and most obvious remnant.

There were in the Edo period other cities whose functions were perhaps more strongly defined than those of the castle town, which tended to combine ordinary commercial and industrial activity with its primary roles of administration and military preparedness. Generally, at the threshold of Japan's modern century, there were at least five other basic types of city, all having origins in early history. These included:

**Ports** (*minato-machi*)

From about 1640 until the 1850s, these, with the exception of Nagasaki, were exclusively for the domestic trade or for fishing, boat-building and like activities. Many prospered, especially in the furtherance of the shipping trade in rice, or from their roles as transport centres in the *sankinkōtai* system. But most had very ancient roots and some, such as the so-called 'free' cities, had flourished on foreign trade and piracy long before the Edo period. Many continued into the modern period, but only a limited number, mostly oriented to the Pacific and Setonaikai (since modern commerce shifted heavily in that direction), were adequate for international trade. Therefore, successful transition and expansion into sizeable cities required either entirely new port construction, or massive reconstruction.

**Post or stage towns** (*shukuba-machi*)

The administrative mechanics of the Edo Government, as described, provided a special thrust to the development of this class of towns, although many had had similar functions previously. Especially important in this process was the system of alternate attendance, which prompted the building (or repair) and maintenance of a highway network centred on Edo, and which, considering the difficulties of travel, required frequent rest stops. Moreover, since *samurai* processions sometimes numbered in the thousands, and there may have been several at a time in any one place, the facilities had, of necessity, to be large and numerous. These settlements were colourful places which featured pens and stables for horses, their equipment and trappings; dormitory-like residences for menials; inns, tea houses and brothels of various classes according to the status of the traveller; and the shop-residences and other buildings of merchants. Towns were often of the *strassendorf* type, with the structures lining the highway so that one had the impression of travelling in a city for long stretches, despite the rather uninterrupted presence of paddy fields immediately behind. The colour and motion of these communities, as well as their

## The establishment of a permanent network of cities: 1573 to 1868

3.3 Kiso Fukushima, an old post town on the Nakasendo (highway) between Nagoya and Tokyo. The town extends along the bank of the Kiso River, an embankment of which can be seen in the foreground. Owing to extreme shortage of land in this mountainous area the town has expanded up the steep slope behind

humour and warmth, were well captured by the artists and writers of the period, as can be seen particularly in the famous and still popular woodblock prints of Hiroshige and his followers. Many of these towns made successful transitions to modern cities, or after 1868 were easily incorporated into conurbations.

### Religious centres (*monzenmachi*)

Morphologically these were often similar to castle towns (although probably more so to those of the pre-Edo period), with a temple (Buddhist) or shrine (Shinto) as the focus of activity, rather than a castle. Many had extremely ancient histories and some in their day, subject to their political backing or even at times to their own military strength, had grown to power. The towns themselves were usually remote, however, and thus were cramped by the physical discontinuities of the site as well as hampered, in so far as their development was concerned, by inaccessibility. In view of this, the town structure also tended to be somewhat more haphazard than at least the Edo period castle town, and the emphasis on amenities for pilgrims that were strongly akin to those of the post towns, and that are still a feature of these places, gave them a different flavour. Generally, these continued after 1868, but with varying success depending on other factors. The most famous today is the city of Nagano (with a 1978 population of 313 197), which is also an important prefectural capital.

### Market towns (*ichiba-machi*)

These were another superannuated association, for which the term 'town' may be inappropriate, since they were merely gatherings for certain lengths of time, of tradespeople and their customers, albeit the meeting times were fixed by custom and the larger had permanent structures. The names of such places, being a combination of the monthly day of gathering, with the suffix *ichi* (market or fair), are an indication of such origins, and the sites are usually located at a convergence of routeways. In the present there are many such names in Japan, but in the main, if a large city has developed, additional functions have accrued through time. The largest of these today is Yokkaichi (Fourth Day Market — 1978 population, 250 396), a burgeoning out-port and oil-refining centre of the Nagoya area.

### Spas (*onsen-machi*)

The common presence of thermal activity in Japan gave rise early in history to the development of spas or local resorts based on this occurrence, and many have become affluent, even though the sources of hot water are no longer natural. Contingent on their locations, which were naturally often distant, growth has or has not been vigorous. Virtually every city of importance today has an association with such a place, and the greatest cities have spawned spas of unusual size. Thus, at present there are such as the Kanto area spas of Atami (1978 population 52 033) and Ito (70 520); the Osaka area satellite of Takarazuka (169 789); the northern Kyushu spa of Beppu (132 386), a kind of national resort; and so on. Many have become amalgamated with a large and adjacent city, and their names have been erased from the roster of places.

In the sense that they later developed into relatively prominent cities, the kinds of towns mentioned above are probably the more important premodern types of concentrated settlement. But it is appropriate as well to mention that there was a class of towns that functioned as a precursor of the manufacturing city. These were numerous, as the Japanese had early become known for the production and even export of certain articles of trade, the reputation of which had been recognised before the 1630s in such places as China and Southeast Asia, as well as at home. After 1600, much of this kind of manufacturing activity became associated with the functions of the castle towns, but much was also centred in separate communities which had, or later attained, fame for their products. Among the items of manufacture were swords and, later, firearms; iron products in general; leather goods and equipment; paper products; textiles and dyed stuffs (mainly of silk and cotton); food and beverages, especially brewed rice liquor (*sake*); and, very widely, tile-making; wood products and the products of skilled carpentry of a wide variety and richness; tools; and pottery. Some of the towns whose products were well known continued into the modern period to become the nuclei of significant cities, and a few are still widely recognised for their ancient specialities. Pottery was particularly noteworthy in this way; the Japanese term *setomono* (porcelain, china, crockery) is taken from the town of Seto, today a flourishing, virtually one-industry city of Aichi prefecture, east of Nagoya (1978 population, 115 458).

## Summary

Although the major groundwork for a hierarchy of Japanese cities in the present was established during the Edo or Tokugawa period (1600–1868), the city as such has roots in very ancient times. The first of importance, both located in the ancestral homeland of the modern Japanese, lying in the lowlands between Lake Biwa and Osaka Bay, were based rather strictly on T'ang period Chinese (hence alien) design, and are represented by Nara (AD 710), although only in its religious monuments, and by Kyoto (AD 794), whose rectangular outline is still obvious, along with its vivid, traditional residences and its famous religious buildings. Perhaps more important than the cities themselves, the Nara (AD 710–84) and Heian (AD 794–1185) periods produced an urbanlike milieu that, although experienced by only a select few, became widely known to the Japanese and later to the world, especially through the literature.

From the end of Heian until the rise, in the sixteenth century, of forces that by the end of the century had led to the unification of political power, cities, except for Kyoto, were generally small and there was little effective integration of their activities. Political and military power resided in castle towns that were mostly inaccessible strongholds, and whose sites usually did not lend themselves to later expansion.

An exception is seen in a few, relatively free port towns that by the

1500s became virtual city-states, the foremost of which was Sakai. Typically, its prosperity was based on foreign trade and piracy and its power was largely monetary, for by supporting key military figures in whom also rested the chief political power, it maintained a kind of semi-autonomy. Since overseas trade was not yet prohibited, these cities often had an international flavour that included Europeans as well as Asians.

The consolidation of political power by Tokugawa Ieyasu and his descendants after 1600, and the rapid growth of their castle city of Edo after 1603 into the focal point of a vast administrative system based on the flow of goods (principally rice) from a network of similar but smaller cities, was the cornerstone of modern urbanisation in Japan and perhaps, at least in size and scope, it was a unique historical development. Rival towns, such as the 'free' cities, were downgraded in importance as the castle town became supreme in authority throughout the land.

The Edo period castle town, thanks to a series of edicts and the absence of war, was located appropriately for subsequent development (as is seen in the large proportion that became modern prefectural capitals); its prosperity was subject to the ability of its hinterland to produce rice and other commodities, and its main function was administrative and commercial rather than strictly military. Therefore the towns were located on flat land and were accessible, via a complex highway system, to other places, and especially to Edo. The system of alternate attendance forced local lords to travel periodically to and from the capital, and the constant traffic promoted not only trade and commerce in general, but the growth and prosperity of cities of all types. Yet development was hampered by restrictions stemming from the strategic suspicions of the regime, which closely watched all travel and refused to permit either the free movement of people, or frequently even a bare minimum of conveniences for the traveller. Travel was therefore slow, expensive and often dangerous.

Despite the handicaps, commerce and trade generally, and cities in particular, grew in size and importance, so that in the seventeenth and eighteenth centuries, Japan may well have been the most urbanised state on earth with a hierarchy of many places of more than 50 000 population, and several very large cities, of which Edo was the greatest.

In addition to the castle town and to certain companion cities designated by the regime as centres of commerce, and on which the proper functioning of the government depended, other classes of cities included: (1) (internal) ports; (2) post or stage towns; (3) religious centres; (4) market towns; (5) spas. There were also many towns devoted to the production of specific commodities and some of these later became modern manufacturing centres. A leading commodity in this way was pottery. All the city types given are represented by modern cities, but often the transition to contemporary prosperity was based on other factors.

# 4
# Changes in the urban landscape after 1868

From the 1850s the feudal regime of the Tokugawa Shogunate was no longer able to withstand foreign and domestic pressures for change and within twenty years the entire structure had collapsed. The reasons for this are extremely complex and many were much longer in the making, but ostensibly the principal moves to end official isolation (and those which ultimately undermined the government and spelled its doom) occurred between Perry's first visit in 1853 and the return to power in 1868 of those who favoured the entry of Japan into the contemporary world.

All aspects of life were altered after this and the milieu presented in Chapter 3 was strongly affected. Most of the communities which had been such important links in the Tokugawa chain of command were deprived of their previous *raisons-d'être* and there was a period of disturbing readjustment. The more prominent eventually continued to grow, while those of less importance, usually of inferior accessibility to the core, lapsed into temporary stagnation and often, in some measure, into decay.

While the Edo period cities, in size and in many of their functions, can probably be considered urban by most means of assessment, they were poorly equipped to cope with the problems engendered by the new national ambitions. Vast reorganisation was needed, and all cities were forced either to become transformed or to be eclipsed in many aspects of development. Generally, urban site characteristics, as has been suggested, were adequate in terms of space and certain other requirements for general modernisation, for the essential geographical components had been uppermost in their selection. But in other ways they were often inadequate, at least without being greatly revamped, and some had shortcomings that called for the construction of entirely new functional entities whose sites were frequently far from ideal for the subsequent spread of the built-up areas. Tokyo, Osaka and other cities, for example, although ports in their own right, had harbours that were too shallow or cramped, or were otherwise unsuited for large-scale international shipping, so that it became necessary to construct outports nearby. Often the sites of the outports, although provided with excellent deepwater facilities capable of being improved by man into first-class harbours, were seriously confined by physical barriers. Kobe exemplifies this, but Yokohama, which grew rapidly after its founding in 1858 from a small local fishing harbour into the nation's most thriving international port, is equally appropriate in many ways.

The inclusion after 1868 of Hokkaido in the mainstream of Japanese development, also led to the building of new cities and soon the growth of Sapporo, Otaru and Asahikawa were noteworthy. These were among the so-called 'pioneer cities', which acted both as spearheads for the settlement of the great island, and initially as defence outposts principally against the Russians in the north. All, of course, eventually became centred on other activities, but none grew from a formerly feudal background as had the majority of the cities of 'Old Japan'. Hakodate, Muroran and Kushiro are somewhat exceptional among Hokkaido cities for having had earlier ties with the ruling elements to the south.[1]

## Some effects of transportation on city growth and change

While other types of cities with unprecedented roots will be considered in due course, it seems appropriate before going further to examine from an evolutionary point of view some of the basic effects of transportation on urban development and change. Castle towns, it has been said, were especially encouraged to expand by routes of travel which before Meiji times, at least for the cities of the core, were mainly in the form of highways. Bulky items (especially rice) were brought to Edo, Osaka and intervening places by sea, and travel by ship was generally common outside the core region, but the Edo period highway network, that bore the strong imprint of much earlier highway development, was the main means for most travellers.

The development of the land routes of 'Old Japan' began in antiquity and evolved along with other cultural attributes, although the natural avenues of travel are so limited that most roads and all important highways have long, colourful and often popular histories.[2] By the time of recorded history the vital parts of the seminal regions of Japanese cultural flowering (as noted in the previous chapter) were linked by long-established arteries whose surfaces were apparently stable enough to permit the use of wheeled conveyances. The two-wheeled bullock cart was extensively used by the aristocracy of the Nara and Heian periods, and ox-drawn trucks hauled charcoal and other commodities long after this in the Kinai lowlands between Lake Biwa and the Setonaikai. Such examples are rare, however, for as other centres of activity grew, particularly toward the northeast, the means of travel became more rudimentary; a condition that saw no important change until after the advent of modernisation in the late nineteenth century.

Between the thirteenth and seventeenth centuries the establishment of strong military and political centres at such farflung locations as Kamakura and Odawara in the northeast, or at Yamaguchi and the Kyushu strongholds in the southwest, led to the increased use of land routes. But while roads generally improved and highway communities formed along important thoroughfares, travel was at the same time slowed considerably by internal upheaval resulting from sporadic internecine warfare which did not end until after the consolidation of power under the Tokugawa banner in the early 1600s.

From about 1620, however, highways were much improved, mainly to accommodate the vast amount of travel by the nobility according to the demands of the aforementioned system of alternate attendance, and the five major highways (*Gokaidō*), together with subsidiary routes, were formed into an extensive system which gradually came to connect the some 265 fiefs of 'Old Japan', or to coordinate with established sea lanes, especially in western Japan. The peculiarities of Edo period travel have been discussed in Chapter 3, but the effect was to create a permanent highway system (later the skeleton for the basic railnet), and, as has been stated, to further the growth not only of castle and stage towns, but of cities in general.

## Transportation after 1868

Following the Meiji Restoration the Government set out almost at once to remove the hindrances to travel and otherwise to encourage public movement. Surveillance barriers were removed from highways and within cities, for example, in the years 1868—69, and by 1873 all restrictions on the mobility of merchants, industrialists and farmers had been lifted. But after careful consideration as one aspect of a series of crucial deliberations that established the future directions of the new Government, it was decided that transportation thenceforth would focus on railways, with highways largely parallel but decidedly secondary. Moreover, with local exceptions, this emphasis has continued virtually until the present, for the construction of entirely new highway systems did not begin until the 1960s and the progress is yet inadequate to accommodate the great proliferation of motor vehicle traffic. Locally, streets were gradually modernised and there was extensive hard-surfacing, particularly in the 1950s, of all national and most prefectural highways, but until recently the ancient highways pattern remained basically unchanged except for railway embellishments.

The building of the basic railnet was a prodigious and costly undertaking which occupied the Government throughout the later years of the nineteenth century and well into the twentieth. British railway experts were initially consulted and many were invited to Japan, not only to plan and supervise pilot construction, but also to man the first trains. Service was inaugurated between Shimbashi in Tokyo and the outport of Yokohama in 1872, and between Osaka and its outport of Kobe in 1874, with an extension to Kyoto in 1877; and by 1880, according to Reischauer, most of the major cities had been linked by Government lines.[3] But also by about 1880 the Japanese had outgrown the need for highly paid foreign specialists, and from that time until the early years of the twentieth century the basic pattern of railroads was slowly forged, with authority firmly in Japanese hands. In 1906 the various segments of the basic railnet were 'nationalised' into a public corporation.

Private railway companies provided important supplementary services, particularly within cities and in places of special need, such as between the cities of the core region. Major construction of private railways as they are known today began in the late nineteenth century with the advent of electric traction and improvements have continued down to the present. Indeed, the growth in efficiency of these facilities for relatively brief passenger and light freight runs has been a vital factor in Japanese economic progress throughout the twentieth century. Some of the private lines have also been converted to 'standard' (4 ft 8½ in) gauge and many are multitracked to permit faster and more comfortable service, but generally the Government lines of the Japanese National Railways (JNR) are of narrow (3 ft 6 in) gauge, mainly thanks to the British, who felt that the rugged terrain, requiring countless tunnels and other difficult engineering works, would militate against standard track width — advice that the Japanese have regretted for more than a century. For although by the time the basic railnet was completed the limitations of a narrow-gauge system in terms of average speeds and general flexibility were appreciated by the Japanese, reconversion to a standard system in the face of other national needs would have been too costly.

The JNR meanwhile has been steadily modernised and is now a huge (though still largely narrow-gauge) system that reaches nearly all parts of the nation and since the 1930s has even provided undersea connections between Kyushu and Honshu. Powered now mainly by diesel or electric energy and despite the continuing dearth of double-tracked lines except where use is heaviest, it has been able to meet with surprising efficiency the constantly growing demands for its services. The scale of this demand, especially for passenger services in the core, has resulted in the construction of a wholly new facility, the now famous 'New Tokaido' line and its recent westward extension in the 'New Sanyo' line, an all-electric, computerised, standard-gauge system with no level crossings, which at present links Tokyo and Fukuoka but which will eventually become the new basic railnet with direct connections by tunnel and bridge to all major islands. The new system boasts the world's fastest scheduled runs, which when added to a record of unusual comfort and safety since the inception of service in 1964 have made it the envy of the world in this field, despite recent public dissatisfaction over excessive noise and vibration which have disturbed particularly those living near the lines. This has resulted in some dampening of enthusiasm as well in the curtailment or slowing of efforts to expand the system. In addition, the petroleum 'shocks' of the early 1970s and the great acceleration of highway traffic, especially of the private kind, have all offered such competition that railways in general have suffered declining revenues. Consequently, costs for passenger and freight travel have risen remarkably and often services have been cut. Notwithstanding, the *Shinkansen* systems remain the most profitable in terms of passenger use in the entire network of the JNR. This public corporation is also constantly experimenting with even more unorthodox means of overland transportation and some of these are hoped to be in practical operation before the year 2000.

Since Meiji times, therefore, the most obvious effects of modes of transportation on the growth and development of cities have come from railroads rather than from other forms of travel. It has only been since about 1950 that buses and trucks have begun to seriously challenge the supremacy of the railroads of Japan, and the influence of local aircraft and the privately operated automobile are even more recent. On the other hand, super-highway facilities have been greatly improved and enlarged, and plans for the incorporation of all major population centres of all islands in a network of ultra-modern toll highways are rapidly becoming a reality. It is obvious that competition will continue to plague the railroads of Japan, especially from alternative overland modes of transport. Meanwhile, another serious competitor is inter-island shipping, which has long been used particularly for low-cost large-volume items, and considering the physical limitations to overland travel, this course also seems destined to continue. Problems of railway freight haulage are being met meanwhile by modernisation, including containerisation and piggyback services operated directly by the JNR, and often the lines employed are the older system, such as the narrow-gauge but double-tracked and electrified Old Tokaido Line, which formerly were devoted to long-haul, relatively high-speed passenger service.

Although the trend was established earlier, some of the more obvious changes in landscape have occurred since the early 1970s as a result of the proliferation of automotive traffic. The old-fashioned two-lane roads and highways which were superficially improved in the 1950s and which remain the foundation of the national highway network, have thus become flanked by countless facilities for the upkeep of car, truck, and bus, and for the well-being and comfort of the rider. Aside from service stations, these include *pachinko* parlours, drive-in restaurants and retail outlets of all descriptions, including supermarkets (in Japanese *sūpā*), and motels. Hence the motorist is often obliged to view the scenic elegance that has always marked cross-country travel in Japan in random fashion, as it still can be seen between clusters of buildings of diverse or bizarre design. Travel by train is more traditional in this sense but the scenery from the roads has all too often been committed to a kind of American-style modernisation.

Within cities, limited access arterial highways, either elevated above or tunnelled beneath existing streets, have been constructed with great vigour, especially in the great cities of the core region. In Tokyo and Osaka (as previously in the building of the Chuo Line of the JNR within the capital), the imperial moat and other waterways have been used as convenient rights of way for arms of the freeway system. On the outskirts of most cities as well, there are new highways to circulate traffic around

## Changes in the urban landscape after 1868

rather than through the city centre, and the term 'bypass' (*baipassu*) is universally applied.

### Further effects of transportation on cities

The establishment of a rail system did not permanently deter the growth of the major cities, for few important places were excluded from the basic

4.1 One of the new toll highways, Chiyoda-ku, Tokyo. The moat of the Imperial Castle is visible on the left

4.2 An overview of the same highway, looking northwest over Tokyo

*Further effects of transportation on cities*

network. None the less, there are examples of areas that experienced stagnation or decline as a result of shortsightedness in releasing land for Government rights of wya. In Okayama prefecture, for example, prosperous coastal towns resisted the railroad for various reasons and the Sanyo trunk line was forced to follow a more difficult, inland route. This was beneficial to the prefectural seat of Okayama and its nearby rival, Kurashiki, but the coastal towns have generally suffered in comparison.

Other towns saw their development temporarily arrested because they were not situated precisely in association with rail lines being constructed, or because their functions were eliminated. The formerly important stage-town of Kawasaki, midway between Tokyo and Yokohama and now a combination Kanto area dormitory town and gigantic industrial city, is an example. Presently the nation's eighth largest city with a population in 1978 of over 1 million, Kawasaki had a total of only 21 391 persons in 1920.[4]

The railroad has been responsible also for a strong morphological effect on Japanese cities in general by causing most modern cities to have at least two nuclei of commercial activity. This came about through a long process stemming directly from the days of initial construction. In most cases, the Government was able only to procure railway rights to land that was adjacent to the focal points of activity. Hence, stations were usually located at some distance from the city centre and often these areas were initially barren of commercial life. In time, of course, business activity developed around the station, together with corridors of such enterprise between the station and the commercial core, and along which local trans-

4.3  Some new high-rise buildings in the Shinjuku district of Tokyo

portation, usually tram lines, were channelled. The station area thus became a place of business and traffic interchange and gradually its position was strengthened until it often came to rival the central business district. Interurban railway lines were also constructed to such station centres, bringing daily commuters, so that the private electric rail terminal–department store complex became a particularly lucrative enterprise in the larger cities, notably in Kanto, Kansai and Nagoya, but mirrored widely elsewhere. A measure of 'healthy' city growth has also been the expansion of access to these station centres from the opposite side of the tracks, first by level crossings and, as traffic pressures have increased and bus lines have overwhelmed tram systems, by means of overpasses, or tunnels.

Underground railway construction began in Tokyo in the 1920s and in Osaka in the 1930s and has progressed extensively in both places, especially since about 1955. Nagoya added a subway system in the 1960s and Sapporo followed suit in the early 1970s, with Yokohama and Kyoto building systems before the end of the decade. Both Kyoto and Kobe meanwhile have subsurface electric lines that are extensions of interurban systems originating in Osaka. In the future subways will become a feature of all cities of at least 800 000 population and probably, as is the case with the newer lines of Tokyo and Osaka, these will run underground only beneath the cities, and will continue as surface lines outside the built-up areas. Subway stations frequently, perhaps as an outgrowth of the complexes of commuter terminus–department store and shopping arcade at certain nodal points in the larger cities and dating from the 1920s, are associated with elaborate and often vast underground malls which may be air-conditioned and are usually provided with an enormous assortment of facilities for shopping, entertainment and other tertiary services, or which may serve merely as protected passageways for pedestrians.

### The functional transformation of historic cities

While the geographical framework of the modern urban pattern had been established by 1868, there were many changes in the functions of cities following the Meiji Restoration. Each type reviewed in Chapter 3 underwent a shift in purpose or, as indicated, was reduced in stature, but the wisdom of the original selection of sites is affirmed by the fact that most of the important cities before 1868 survived the process of transformation.

#### Castle towns

The more important of these were almost immediately able to resume a position of leadership. For a short time, however, it appeared that they would suffer a loss of function, as the props to their security in the form of subsidies and other protection by feudal decree had suddenly been removed. On the other hand, since castle towns generally occupied the

## The functional transformation of historic cities

more propitious places for city growth, this condition was a strong factor in tiding them over a period of difficulty. Free travel brought them increasing trade over roads and later railroads designed to provide maximum access by using to advantage the best features of an irregular terrain. Greater trade led to such service functions as finance and banking and, along with these, industrial development often initially based on local products for which they had long been famous. Light industries such as food processing, spinning and weaving, *sake* (brewed rice liquor) and *shoyu* (soy sauce) brewing, clothing and farm implement manufacture, became typical of the smaller castle towns, while the larger added perhaps some such heavier industry as metal working, shipbuilding, rolling stock and eventually automotive equipment manufacture, or in the realm of brewing, since it came to be made by large, consolidated concerns, the production of beer. Most of the sites, as indicated, were also logical points from which to exercise control over communications. The main castle towns therefore became the natural foci of the telegraph, telephone, transportation, radio and later television networks.

Culturally, the leading castle towns were or grew into educational centres, although until after the Second World War, when prefectural universities were established, the education they provided was limited to the high school level except for technical schools, certain teacher training institutions and five Imperial universities established between 1889 and 1918 at Tokyo, Kyoto, Sendai (Tohoku), Fukuoka (Kyushu) and Sapporo (Hokkaido). The educational system, however, was complex and it is difficult to draw exact comparisons with that of other cultures. Many of the old higher schools, for example, were perhaps of junior college calibre, as may be affirmed by their elevation after 1945 to university status. Otherwise, higher education was (and largely remains) heavily concentrated in Tokyo, which has a vast hierarchy of such institutions, public and private. So dense is the concentration of educational facilities, in fact, that at least one major public university (now known as Tsukuba National University but formerly Tokyo Kyoiku University) has been moved to an entirely new and hitherto rural location. Staggering pressures on public transportation and other Tokyo services, virtually throughout each day of normal activity, would seem to justify such action, and it is possible that in decades to come there will be major movement of academic life away from the larger cities, Tokyo in particular, thus reversing in some measure the tradition of the castle town as an educational centre, which began long before the Meiji Restoration.

After 1868 the castle towns that were designated prefectural capitals also became known for other educational, cultural and service functions including medical facilities and training, printing and publishing (especially of local newspapers) and, in general, for public enlightenment. A usual feature of these cities is the civic centre containing a variety of public buildings such as lecture and concert halls, museums and general use auditoriums, as well as the city and prefectural office buildings in which certain of these may be housed. Today, partly because they were destroyed or so heavily damaged during the war, these buildings are often of striking design and furbishment and the whole area is attractively landscaped. As seats of local government, prefectural capitals before 1945 were frequently army garrisons, which helps to account for their almost universal destruction by allied aircraft in the last summer of the war.

As centres for amusement and recreation, each prefectural capital also developed a theatrical section and, at least before the banning of prostitution in 1958, an open and distinctive red light district. In the present, restaurants and coffee houses, *pachinko* (pin-ball) parlours and bowling alleys abound throughout most prefectural capitals; there is characteristically a strong concentration of these in the central business district or the station area, while the back streets of these sections are usually devoted to bars and other business establishments. In the outskirts of such cities there are usually larger structures in this category, including the inevitable baseball stadium, race tracks and zoological gardens. Generally paralleling the trends in other of the world's industrialised nations, Japanese urban growth has been characterised by a relatively rapid increase in tertiary and secondary industrial activity, but there has been a marked decline in the primary sector, and this is particularly true of the prefectural capitals. Daily commuting is therefore a distinct feature of life, and the increase in such centripetal movement from an expanding suburbia has been a noteworthy tendency in recent years. Thus, in these and other capacities, either added after the Restoration or based on long-established specialities, the castle town turned prefectural capital has become a truly multifunctioning regional centre, with variety as the keynote in (mainly light) industry and the services, rather than specialisation.

There are, of course, many ex-castle towns of lesser dimensions, so that in the writer's opinion there are two categories of modern cities with such background. One is the general run of prefectural capital, of which thirty-three of the forty-seven are of this lineage. The other is a larger group of cities which were not given major administrative duties by the central government after 1868, and are therefore of generally less importance in the present urban hierarchy. A few have managed to acquire some attribute that has led to subsequent development, but usually this has happened because of factors which may not be directly related to their origins as castle towns. Fortuitous placement in or near the modern urban core has spelled success for some, while others owe their affluence to the addition of secondary or tertiary industrial specialities. Hamamatsu, for example, is well placed within the Tokaido region and has also become a centre for the manufacture of pianos and motor vehicles. Hakodate, a former castle town, oddly located far from the core in southern Hokkaido, is the main entrepôt for the northern island and is also a successful port for fishing and other sundry activities, while Himeji, whose magnificent White Heron castle is a famous feudal remnant, commands a relatively broad plain west of the Kansai area and has developed many important local functions, including transportation, industrial production and tourism. There are other flourishing ex-castle towns that have not become prefectural seats but the list is not long, for the advantages accrued mostly to those that had political importance after the Meiji Restoration.

**Port cities**

In the Edo period, except for Nagasaki and a small number of harbours in that area, these were entirely for domestic traffic. After the Restoration, some survived as separate entities by undergoing extensive face-lifting, but many were erased from the roster of city names by absorption into larger, formerly inland cities. Many more remained through the modern period as small, rather isolated ports devoted to local fishing, boat construction, seaweed gathering and curing, pearl culture and so on. Those that survived as major cities were able to add important manufacturing and service functions. Nagasaki, for example, is a vital modern port, but is also a major industrial and administrative centre, and the same can be said for Osaka, Niigata, Shimonoseki, Nagoya and others, although obviously many of these owe their success not only to their roles as ports, but also because of their status as leading Edo period castle towns whose political functions were continued after 1868. Thirty-four of the sixty-nine cities with populations of more than 250 000 in 1978 were ports.

**Stage towns**

The abolition of the system of alternate attendance for the nobility, which

4.4 Apartment and single-family dwellings in Sendai, August 1980

came only six years before the Meiji Restoration, put an end to the principal function of the post or stage towns, and the coming of the railroads in the 1880s and 1890s established the pattern either for their development or decline. Since the railways were built to follow the old highway systems, certain former stage towns became railroad centres and in time added minor industry and other functions. Perhaps in more recent years with the advent of long-distance trucking and bus services there has been some revival of their initial activity. On the other hand, modern highway culture in Japan, as elsewhere, tends toward random growth along new or modernised highways rather than to clustering in historic associations. Of the sixty-nine cities in 1978 with populations greater than 250 000, few owe their origins entirely to this function. In most cases there were combinations such as castle—stage or castle—port—stage, or temple—stage towns. However, several, including Sendai, Kawasaki, Hamamatsu, Utsunomiya and Nagano, had strong roles as stage towns. Only Urawa, now capital of Saitama Prefecture and a Tokyo area satellite, has descended wholly from such origins.

### Religious centres

By their very nature, the religious centres retained much of their vitality after 1868 and often increased their standing with the removal of travel restrictions. The fortunes of religious sects, whether Buddhist, Shinto or unconventional, have tended to vary through the years, however, and these vagaries have been important factors in the growth of communities that are dominated by such activity. The more fortunate are those that have been able to add other functions, especially tourism, but in some cases industry, as mentioned in Chapter 3 in the case of Nagano, the prefectural capital, which has become typically multifunctional. According to the religious beliefs of the sect involved, there is often a festive atmosphere in this type of city, with amusements playing an important part, especially at times of celebration. In this regard, religious centres are similar to resort towns, even in respect to the kinds of amenities that are offered, and the site characteristics are also often comparable, at least for hot springs resorts. One rather unique town in this category which attained city status in 1954 is the city of Tenri, headquarters of the rich and populous Tenri sect of the so-called popular Shinto order. Located about 10 kilometres south of Nara, Tenri in 1978 had a population of 59 959.

### Resorts

The designation of this type of city has been changed from hot springs or spas, as in Chapter 3, to account for the rise of towns which had no historic roots. Usually, however, the more affluent modern resorts have grown out of ancient spas, even though they serve other functions of a recreational nature. Beppu and Atami, as mentioned, are still essentially spas and are the most famous resorts in Japan, but there are well-known places which cater to those who sail, swim and ski, although generally few have attained city status. On the whole, resorts are usually found in association with larger cities. Beginning in the nineteenth century small, interurban electric lines usually connected the resorts with the JNR station of the parent city, and today these connections have been modernised to accommodate a large amount of bus and automobile traffic and in consequence of general prosperity and improved access the resorts have grown. Some, such as the ancient spa of Dogo, have been joined to a larger city. One of the most recent places to have acquired a recreational function, and to have been transformed, is the Senri area, northeast of Osaka and the site of EXPO '70. Until the exhibition grounds were constructed this was an attractive hilly region partly used for suburban housing, partly bamboo forest. Commuters to Osaka, only 10 kilometres away, used the Senri line of the Hankyu electric railway, but with the development of the area, transportation access was immensely widened to include spurs of the municipal subway system and other private railways as well as super highways. The formerly serene atmosphere has been entirely superseded within a few years by the most modern attributes of megalopolitan living and it seems certain that the area will continue to have functions far different from those of the past. Close to the fairground also is the huge Senri 'new town', one of Japan's first planned housing areas and the forerunner of many such developments, not only in the Tokaido Megalopolis, but rather generally.

During the Edo period such Chapter 3 categories as 'free' cities and market towns (*ichiba*) became outmoded by the socioeconomic system that developed under the Tokugawa rule, but after the Restoration and in the ensuing years new kinds of cities have arisen. Outports and the new cities of Hokkaido have been discussed, as have those which were famous for particular handicrafts or products and have profited by such specialisation. But obviously modernisation and particularly industrialisation have called for the creation of other categories. Among these, and of which examples were given in Chapter 1, are cities whose basis is industrial and whose locations, initially at least, were affected by nearness to the source of raw materials. Growing out of these, and from the general processes of urbanisation, wherever a significant proportion of the work force has sought new housing and as rising incomes have permitted satellite communities have sprouted. Until the 1950s these were limited in size and in distance from the place of work, although there had been sizeable growth of suburban culture, especially around Osaka and Tokyo, since the period of economic stimulation during and after the First World War, when private electric lines first began to effectively inflate the living spaces of the larger cities. Beginning in the late 1950s, however, new and vastly larger housing complexes have developed, first around the largest cities and in recent years rather universally, with the typical dwelling unit being an apartment, despite the thousands of small house lots that are currently

4.5 View of new large-scale *danchi*, August 1980. Tama New Town, in the western suburbs of Tokyo

being fashioned in the same general areas. Commuting distances have also increased greatly around the big cities, thanks to more rapid and in some cases new electric trains and especially to highway improvements. Some of these dwelling compounds (*danchi*), or 'housing estates', such as the aforementioned Senri 'new town', are combinations of single-family dwellings and large apartments. These are linked to city centres by rapid transit lines or by bus routes, and in some there are local facilities for shopping and other services (with these features tending to become more elaborate according to the size of the city and the newness of construction), but there has been little development of secondary industry, at least as an integral part of the enterprise. *Danchi* may be constructed by the Public Housing Corporation or by private agencies, and many have grown from a combination of such forces.

Other types of modern communities are the defence cities, mostly naval and air, with the former having attained very significant proportions, both before 1945 and with strong American influence in the postwar period. Yokosuka, with a 1978 population of 406 985 and Sasebo with 254 480, are among the leading cities in size and vitality. Important air bases are often too close to larger cities to have become urban centres in their own right, but Iwakuni of Yamaguchi prefecture, with a 1978 population of 111 395, owes much of its prosperity to a largely American air facility, as do Kanoya (69 880) of southern Kyushu, and Wakkanai (54 945) of far northern Hokkaido. Some students of Japanese urbanisation have listed traffic cities as a separate category, and although these have been discussed briefly in the foregoing under stage towns, it appears that in the future as the new railway and highway networks proliferate, there will be marked influence on the development of former and perhaps entirely new cities of this category.[5] Otherwise, although this type has yet to be realised, the future is also bound to see a class of cities whose principal business is *education*.

## Summary

Following Admiral Perry's visit to Japan in 1853 the 250-year-old Tokugawa regime, which had followed a policy of seclusion and the proscription of much human activity, fell from power and by 1868 a new age, no less restrictive in many ways, was born. Most cities experienced a difficult period of adjustment to unprecedented conditions, but the major ones, castle towns especially, survived and continued to grow. The main castle towns were best able to serve as regional capitals because they had usually been carefully sited, and the original requirements were still sufficient, at least in terms of space.

Outport construction was necessary in some places to allow international shipping, but the new sites, exemplified by Kobe and Yokohama, were often physically hampering to large-scale expansion, which usually occurred in any case.

Hokkaido cities, with minor exceptions, were all new in that they had had no previous ties with 'Old Japan'. As such, they were mainly pioneer cities which functioned initially as focal points of settlement or as military outposts against the Russians on Sakhalin to the north.

Cities were generally much influenced in location and growth by means of transportation, which before the Restoration meant highways and coastal seaways, both having roots in earliest history, but which became particularly well developed during the Edo period. It was then, for example, that a network of highways, supplemented by established sea routes, was first integrated into an effective system for promoting economic and urban growth on a national scale. So limited were the natural lines of communication, however, that the feudal highway network later became the basic pattern for the modern railroad matrix.

Travel after 1868 was actively fostered by the Government and attention was centred on railroads, which became generally efficient and pervasive, despite such limitations as a dearth of double-tracked lines and a narrow gauge system, except for certain private interurban lines and the latest Government projects. Highways, until the 1950s, were largely unimproved, but until then existing facilities had served the nation remarkably well and were instrumental in the general growth of all major cities.

Private railways also helped meet demands for rail service, especially within and between the largest cities, and these were often modern in the extreme. In recent years, the Government has invested heavily in an entirely new superspeed rail network, the first segments of which have enjoyed efficient but not problem-free operation since 1964. Eventually, this will become a new basic railnet, with bridge and tunnel connections to all main islands. Meanwhile, the older though still modern lines, which formerly emphasised passenger service, will be gradually relegated to modernised freight transport, except for local runs.

Within the past two decades highways have been much improved. The older roads, although for the most part hard-surfaced, are generally too narrow or limited in extent to handle the enormous volume of traffic. Therefore, a completely new system of modern, limited access, toll highways is in construction which will eventually have a national coverage similar to that of the JNR's future railnet. Limited access arterial highways within Tokyo and Osaka and to a lesser extent in Nagoya, Kyoto, Kobe and other cities, have been elevated above or tunnelled beneath existing streets in an attempt to speed the circulation of traffic, and it is clear that such construction will become more general in the years ahead.

Most cities benefited by railways and are currently experiencing the stimulation (or stagnation) of highway traffic, supplemented by domestic airlines. But the major influences on city development have come from railways, with the station complex often growing to rival the central business districts. Subways date from the 1920s in Tokyo and from the 1930s in Osaka, and are now much augmented and modernised in both, indicating that in the future intracity mass transportation for large cities will be mostly underground. Lavish subsurface shopping and amusement complexes have often been built along with major subway stations.

All towns underwent functional changes with the castle town generally becoming a regional centre of two kinds: those with major administrative functions and those without. The former are the leading cities of the country and make up nearly half the forty-seven prefectural capitals, which generally feature light secondary and such vital tertiary industry as government, finance and banking, communications and the services in general. Education is particularly important in prefectural centres, with Tokyo being so top heavy in this regard that there are signs of a major decentralisation of this function, and the creation of a new class of educational centres.

Ports have grown in importance only if they have been able to accommodate modern shipping, or if other functions are noteworthy. Many significant castle towns, for example, are also prominent ports. Otherwise, ports, though numerous, have mainly local functions.

Stage towns have also had to shift with the times, the most successful being those which have added various modern functions. Few owe their present position to their former activity, but some have become railroad centres.

Religious centres profited by increased public mobility, but their progress has been heavily subject to general economic conditions or to variations in the fortunes of the particular sects involved.

Resorts are based generally on former spas, but there are new kinds catering to more modern forms of recreation. A few have become major cities with a national or even international clientele, but generally their prosperity has depended on that of the city with which they are most closely associated.

Since 1960 industrial cities and satellites have grown in importance along with the surging economy, and though much of this development has occurred in the urban core, the movement has been felt even more widely. Recently, satellites especially have tended to proliferate into huge 'housing estates', or into rambling, homogeneous suburban tracts, and the phenomenon has begun to grow characteristic of urban expansion, regardless of location.

Defence needs have influenced, somewhat narrowly since military development has been heavily curtailed since the Second World War, the growth of cities in widespread locations, and based mostly on naval and air operations, but there are several cities of important size in this category.

# 5
# Historical aspects of the commercial landscape

The proliferation of small manufacturing based on traditional crafts during the Edo period is such a prominent stage in Japanese economic history that one is tempted to ignore the considerable growth that occurred well before the 1630s in conjunction with foreign trade, particularly after the fourteenth century.

A pattern of exchange with China and Korea was established early, before the Japanese had attained significant experience of the high seas, in order to provide commodities that were not yet produced locally. For example, although the Japanese had been expert metallurgists and metalworkers since prehistoric times, substantial quantities of iron were being imported as late as the seventh century for forging into tools, utensils, and especially for swords.[1] These last were to become a speciality of great importance not only to the export trade but to the whole advancement of Japanese technology. However, by early Heian times, about a century after the iron imports here attested, the exploitation of various small and scattered deposits in Japan had led to the moderate production of iron goods, eliminating for the time being the need for imported raw materials.

Another early import was Chinese currency or copper 'cash', which became increasingly needed as foreign trade grew. Although the Japanese had attempted as early as the eighth century to popularise the use of money for foreign exchange, their own techniques of minting were inadequate until the time of Hideyoshi in the sixteenth century.

From the twelfth century to the early years of the seventeenth, Japan was racked by internecine warfare which symbolised the transformation of cultural values from the effete tendencies of the preceding Heian age to the ascendence among the elite — and to a large extent generally — of the martial tradition. During this time foreign trade expanded, not only with China but throughout East Asia and even further afield to many parts of Southeast Asia. It must be allowed, however, that not all the participants were Japanese, nor were all their efforts ascribable to trade in the usual sense. Privateering had become so common and was so particularly associated with the Japanese, who used military power to encourage commerce, that the Chinese, out of constant embarrassment at the depredations of Japanese pirates, especially along their heavily indented southeastern littoral, were frequently obliged to pursue trade more or less against their better judgment.[2]

Much earlier, indeed by the Heian period, the genius of Japanese artisans had achieved a blending of imported styles and techniques which was translated into unprecedented creativity. As time went on the products of this genius grew in quality, number and individuality. Moreover, the distinction between aesthetic and practical art was much blurred. Useful items of lacquer, textiles, ceramics and porcelain, leather, wood, metals and other substances were frequently crafted with such care and artistry that even the most pedestrian have long been considered *objets d'art* of the first rank. In Japan and throughout the world, many have become museum treasures or are privately coveted by collectors or as family heirlooms. Even in the present, the folk art tradition, based on this early flowering and subsequent evolution, is a vital feature of Japanese culture, albeit current hand labour shortages may be reflected in the clever use of machines to fashion the products. On the other hand, the perpetuation of the true tradition is probably furthered by the contemporary practice of officially bestowing on certain renowned artisans the title 'living national treasure'.

It may be too much to suggest that handicrafts of such excellence were direct precursors of modern industry, but at least two considerations seem to relate them to such development. One is the technical skill acquired through centuries of devotion to the perfection of a single object. Manufacturing was usually performed by families, who, from the fourteenth century to the seventeenth, were placed under the aegis of guild organisations (*za*). Consequently production was constantly standardised and improved. There are many possible examples, but especially relevant in this regard are textiles, notably silk, the machine production of which was one of the first and most rewarding of the modern industrial exports of the late nineteenth and early twentieth centuries. Another is metallurgy, particularly the making of swords but not excluding other military gear and a large assortment of cast and forged items for ordinary household use. The second factor which may relate history to the growth of modern industrial practice is the *scale* of manufacturing activity in feudal Japan. Long before the Edo period Japanese entrepreneurs and their agents had managed to create a large and enthusiastic market for such products as swords, halberds and other weapons, painted folding fans (a Japanese innovation), and lacquerware. Equally important, Japanese artisans appear to have developed the ability to satisfy a growing foreign and domestic demand. A large proportion of exports consisted of manufactured goods, but also included were such basic raw materials as copper and sulphur, with which the Japanese Archipelago is unusually well endowed. Foremost among the export items were swords of such matchless quality as to have inspired, in subsequent reviews of the period, laudatory comment of which the following is typical.

Technical progress in sword making was such that, for strength and edge, the Japanese sword of the thirteenth century, if not even earlier, excels the work of all other makers in whatever country, before or since. It was a definitely original contribution by Japan to the applied arts.[3]

The specialisation in the manufacture of superbly tooled and decorated swords transcends the conventional concepts of industry, manufacturing and handicraft, for the practice incorporated all these and, in addition, the finest elements of the Japanese artistic tradition. Furthermore, the swordsmith's art was imbued with a martial spirit that came to flower in the early years of the feudal age. This mystique, with its code of loyalty, valour and stoicism, and its inordinate emphasis on the sacredness of the sword (as one of the three legendary imperial symbols, along with the mirror and the jewel), was at its height in the twelfth and thirteenth centuries: it was also closely associated with Zen Buddhism, and it is no wonder that in the fourteenth and fifteenth centuries, when the scope of trading was expanded and the Japanese sword became a favourite export article, Zen monasteries were often important backers of such commerce.

The best-known swordsmiths were often given honorary rank or official titles, so highly prized was their calling and so important was it to the fortunes of the military aristocrats. The processes of sword production, whether the forging of the blade, the various stages of blade decoration, or of producing the infinite embellishment of the mount, most of which tasks were performed by separate specialists whose expertness was derived from generations of practice and the handing down of technical details from father to son, were universally intricate and painstaking.[4] Even the forging of the blade, which ordinarily might be conceived as a relatively simple process, was, in the hands of the twelfth- and thirteenth-century Japanese swordsmiths, a vastly complicated and time-consuming series of operations.

Thus, sword-making and the business of producing armour generally was accorded an unusual status in a profession that otherwise might have been of low social standing. The guild organisations which developed in mid and late feudal times, in fact, were instruments to protect the artisan against ruinous tolls, cut-throat competition and miscellaneous harassment by those of higher social position, which included the bulk of the citizenry. Other artisans who worked to satisfy more ordinary and chiefly domestic needs, despite a devotion which was equal to that of the armsmakers and regardless of the often exquisite nature of their products, many of which have since attained immortality, were not so deferentially treated. None the less, the names of the better artists and artisans of history are generally and even widely known down to the present. The aesthetic appreciation of beautiful objects and of their creators remains a strong characteristic of the individual in Japanese culture, often regardless of economic standing or educational background.

In spite of the very specialised nature of much manufacturing in feudal

Japan, as exemplified by the professions of the swordsmith and armourer, whose methods were often of such meticulousness as to be laboriously slow, a surprisingly large number of articles was made available for export. This attests, as mentioned, both to the size of the foreign market and to the capabilities of the producer, not to mention the organisational abilities of those who assembled the goods and arranged their shipment. In the year 1483 alone, as many as 37 000 swords were exported to China where fine Japanese swords were in such demand as to bring up to 10 000 copper cash apiece.[5]

Overseas trade and piracy were fostered in Japan by several powerful elements. In addition to the Zen Buddhist organisers mentioned above, and as feudalism became more deeply entrenched, commercial expeditions came to be sponsored by certain *daimyō*. As a result, facilities for trade were improved within Japan. These included the enlargement of ports and harbours, and, as seen in Chapter 3, the growth of commerce was so vigorous that cities such as Sakai and Hakata became rich, populous and for a time even semi-autonomous.

In their dealings with the Chinese, the Japanese were not so insensitive as to rely wholly on swashbuckling in the promotion of trade, but as trade with China normally entailed the adoption by the lesser party of an official attitude of subservience, attempts by the Japanese to maintain this attitude were seriously hampered by several factors. For one, there was at this time no permanently powerful central authority, or if occasionally the *shōgun* was a strong figure (as occurred in 1411), he might suddenly, out of a surge of nationalistic feeling, simply cancel existing agreements. For another, the Chinese were openly unenthusiastic about commercial exchange, especially with those who failed to accept a tributary relationship. As a result there was usually an increase in unofficial or illegal private trade and in outright piracy. On the whole, however, in the widening arena of Japanese maritime activity prior to the seventeenth century, commercial operations were increasingly vigorous.[6]

## Commerce and trade in the Edo period

The exporting of Japanese manufactured goods and raw materials and the importing of foreign commodities continued after 1600 and such exchange was important in the first years of the Tokugawa period. Especially between 1600 and the edicts of 1633 there was a marked spread of commerce and industry as initially business was actively encouraged. Steps were taken, as described in Chapter 3, to improve transportation and there were other progressive changes such as efforts to remove the increasingly restrictive influences of the old guild organisations. There was also movement in the direction of monetary reform and in the standardisation of weights and measures; and, in line with policies that had been favoured by Hideyoshi and his predecessors, the regime even encouraged foreign trade. As cities became larger and more essential in the basic economy, their inaugural prosperity being greatly aided by the *sankinkōtai* system which led to a proliferation of highway traffic, there was notable growth of the service industries, paralleled by a spread of secondary (chiefly cottage-type) enterprise. The evolution of commercial life throughout the Edo period was not, to be sure, such as to produce an industrial revolution comparable to that of northwestern Europe after the late eighteenth century, but considering the gradually expanding efforts of the shogunate to regulate all human activity and to maintain at all costs the *status quo*, there was surprising growth, especially at the beginning.

But as mentioned, the governing policies of the Tokugawa *shōgun*, which aimed to complete the unification of the country, were based on a delicate balance of military and political forces which was controlled more by a combination of traditional loyalties and intimidation than by outright force, which the regime could never entirely muster. Any possible challenge to the central authority was viewed with great suspicion. Therefore, after about 1615, external commerce, which might have brought disruptive elements in its train, slowly came under tight restrictions; from the early 1630s it was virtually eliminated except in and around Nagasaki in far western Kyushu, and even then it was subject to the most rigid control. A few may have risked death to pursue illegal trade, but such was probably very minor and in any case, there would be little reliable verification. Christianity, which by the early seventeenth century had gained a strong foothold in western Japan because past Japanese leaders in their desire for foreign trade had considered religion a necessary concomitant of such interchange with the West, was also ruthlessly eradicated. The very desire for foreign trade was likewise strictly inhibited, although trade in some measure with the Chinese and Dutch, together with the closely regulated introduction of outside influences, continued throughout the Edo period.

The development of Japanese commercial and industrial life between the 1630s and the coming of the Perry expedition in 1853 was therefore overwhelmingly a domestic process. It is doubtful whether foreign influence was brought to bear in any significant way on the manufacturing processes until the very end of this long period, although there was some in the making of the firearms and small cannon which were in common use from the onset of the period. Some foreign influence is also seen in construction techniques, especially of castles. The Japanese thus generally failed to benefit from most of the scientific and mechanical advances of the West between the early seventeenth century and the mid-nineteenth. There was, of course, much curiosity, and a certain amount of concrete information (especially in the field of medicine) was made available to a select few through translations chiefly of Dutch writings, and the Dutch ultimately became the leading purveyors of Western knowledge in Tokugawa Japan. Late in the Edo period there were also some individual and usually isolated attempts to apply imported information, mostly by comparatively affluent and forward-thinking *daimyō*, and since the bulk of such learning was acquired in the vicinity of Nagasaki, it is not surprising that efforts to establish the first textile mill, the first modern iron foundry, and to produce a miniature prototype of a steam locomotive and other railroad equipment were made in parts of Kyushu.

But what the Japanese lacked in terms of stimulation through free association with other cultures they may in part have gained in the development, through their own resourcefulness, of more efficient and often unique mechanisms to guide their evolving commercial society. Reischauer points out that the political stability of the time, one of the longest enjoyed by any comparatively recent culture, was conducive to the growth of commerce and industry, although as the Edo period progressed this was certainly not the intention of the regime.[7] The very arrangement of social classes, with the *samurai* at the top, followed by the farmers, the artisans, and finally by the merchants, as the only 'legitimate' groups, made commercial activity difficult.

The Edo regime placed its main economic emphasis on the development of irrigated rice agriculture, which within the technological limits of the time was made to produce to the fullest and which, as seen in Chapter 2, registered significant advances in various directions. Indeed, it is likely that agriculture in Edo period Japan was technologically superior to that in any other culture of that day. But at the same time as agriculture was achieving its greatest development, it was also being required to provide virtually all the revenue. Moreover, it was during this same period, as stated earlier, that the burden of revenue production became progressively heavier as the number of unproductive citizens multiplied. Surplus rural population, reflecting general population increases attributable to such factors as the absence of war and advances in agricultural technology which caused initial increases in the food supply, flocked by the thousands to the cities, especially to those cities which now constitute the national urban core. In order to ease growing economic hardship, farm families, in what spare time they had, turned to the production of local handicrafts and other items of manufacture, thus providing a species of precedent for the contemporary practice of part-time farming. These local specialities were later organised into fief (*han*) monopolies, in part to help bolster the sagging fortunes of the local elite as they became increasingly indebted to the rising merchant class. Impecunious *samurai* saw in local industry an opportunity to squeeze revenue from an economic sector other than agriculture. In the main, however, the industrial and commercial sectors were taxed much less heavily than the agricultural, and consequently enjoyed a substantial advantage.

It cannot be said, therefore, that industrial development in Edo times was such as to provide a solid base for later industrial modernisation. Yet cottage industry increased after 1868 and grew to the extent of creating what has often been termed a dual industrial structure, a characteristic of Japanese economic life that will be discussed in Chapter 7. Rather, Edo period developments were important in such other ways as the creation of

*Historical aspects of the commercial landscape*

5.1 Used-car dealer on the Sendai by-pass, August 1980

a unified national market, and the formation and growth of commercial firms which bore the strong and characteristically Japanese imprint of paternalism, a quality that to many is still a perplexing feature of Japan's industrial structure. Valuable experience was gained in the whole area of commercial and industrial exchange, and this led ultimately to the establishment of prominent business families, some of whom became famous as *zaibatsu*, whose operations came to embrace virtually every phase of production and distribution. Certain of the now internationally known Japanese names in business, finance, industry and trade, including such giant corporations as Mitsui and Sumitomo, owe much of their success to the early Edo period struggles of their founding fathers who by strength, cleverness and perseverance were able to overcome popular as well as official prejudice and eventually to triumph in an essentially hostile world. Many were able (once they had become economically important) to elevate their social positions by making advantageous marriage alliances with aristocratic families whose fortunes were dwindling.

As the Edo period progressed and as changing economic circumstances slowly forced the relaxation of official restrictions against such activity, individuals in business and industry formed themselves into large associations (*kabu-nakama*) which ultimately became power blocs, the fortunes of which tended to rise and fall with the changing attitudes of the regime. Particularly after 1721 when the *shōgun* Yoshimune removed the final prohibitions against such organisations and these associations began to enjoy almost official sanction, they were instrumental in stabilising prices, fixing interest rates and establishing monopolistic controls, or at least in attempting to bring about such reforms. Industry, especially where there was specialisation, such as among the textile weavers of Kyoto and the pottery makers of northern Kyushu, also came to be organised into associations, and there were even such arrangements among certain service workers, typified by building contractors and carpenters. About midway in the Edo period there were more than 100 professional associations of various kinds in both Osaka and Edo.[8] Again, the vast scope of activity is an indication of development and of its importance in introducing an atmosphere of changing values and new modes of behaviour, at least on the part of a relatively large, active and influential segment of the public. An appreciation of such background may help to qualify in some way the commonly held belief that the extremely rapid commercial and industrial development that took place after 1868 was generally unexpected and is therefore such as to inspire wonder and surprise among observers.

### Edo period industry

However, whatever industrial advances were made during the Edo period, were usually made at a heavy price, and all were perhaps less successful than they might have been had the regime's powers of centralisation been stronger. Sheldon gives the following summary of the contradictions in this regard:

There were three factors in particular which militated against an in-

5.2 Drive-in restaurant, Kanagawa Ken, August 1980

crease in merchant influence and independence: the tendency for the feudal ruling classes to bring commercial activities under their control so far as they could; the existence of provincial economies which opposed the development of a nation-wide economy, and the exclusion policy which barred the path to economic power through foreign trade — a path which in Europe by 1600 had already proved of such importance in the rise of the merchant class.[9]

Always therefore there were important moments of retrogression when progressive action to further industry and commerce was stifled by the conservative ruling *samurai* class whose strength, though it wavered from time to time, at least until the time of Perry, was without significant challenge.

In the absence of any meaningful foreign trade, industry in the Edo period was aimed almost exclusively at the home market. As such, it was concerned mainly with the production of such necessities as items of food and drink, textiles, household goods of various kinds and materials, tools and utensils, articles for building and construction, and military equipment. Craftsmanship continued to be of surpassing excellence and the scope of manufacturing was greatly widened according to the general growth of commerce in both urban and rural Japan. There was large-scale production of often exquisite articles of wood, lacquer, silk, cotton and other textiles, fibres, stone and glass, metals and gems, ceramics and pottery, and other materials. Techniques of cabinet-making and carpentry, masonry, metallurgy and other construction processes were especially well developed. Architecture and allied decorative arts, stemming from a distinguished heritage, produced not only fine examples of official and public buildings, but a whole style of construction for virtually every economic level and incorporating such innovations as prestructuring and prefabrication.

Generally, certain artisans tended to be located at random while others, such as the makers of some large-scale commodities, of which Kyoto textiles and northern Kyushu pottery have already been mentioned, often came to be clustered in specific areas that grew famous for a particular product. The localisation of the textile industry in the Kyoto Basin, for example, is in large measure explained by the quality of the water in the streams that flow through the city, which was (and still is) especially suited to the processes of setting colour in the cloth. The suitability of water is also important in the major localisation of the *sake*-brewing industry as exemplified in and around the town of Nada, east of Kobe.

The important occurrence of raw materials often led to the concentration of industry, even in Edo times, as seen in the centralisation of pottery production based on local clay deposits and fuel in the Nagoya area. But equal in importance to natural substances and their influences on the growth of industry were the human and economic effects of local specialisation, which tended to attract a coterie of professionals, including tech-

5.3 People boarding minibus to attend driving school, near Tama New Town, Tokyo suburbs, September 1980

nicians, not only in production but also in the business of wholesaling and transporting. Large accumulations of labour were also instrumental in establishing new and important settlements, although generally, aside from certain pottery towns, few grew into important cities until after the Meiji Restoration.

Reischauer points also to the differing nature of manufacturing according to political affiliations and degree of urbanisation. The *shōgun*'s domains, which constituted much of the contemporary urban-industrial core area, had more than their share of industrial and subsidiary activity, while the outlying areas — generally of lower population and economic development — had proportionately less.[10] Paradoxical as it may seem, moreover, the ratio of *samurai* to commoners within this same domain (Edo, Kyoto, Osaka and environs) was far less than in outlying *han*, a circumstance which allowed the townsmen (*chōnin*) and other citizens greater freedom from official restrictions on commercial activity. As mentioned in Chapter 3, the great city of Osaka was almost wholly a centre of commerce, earning for its citizens a somewhat unsavoury reputation for business acumen that has persisted down to the present.

Urban commercial development in what is now the core region reached its zenith in the so-called Genroku era (1688–1704), slackening thereafter in favour of increasing activity in certain parts of the provinces, especially as checks on freedom of commercial growth were imposed in the cities.

## Historical aspects of the commercial landscape

The associations of merchants and industrialists, for example, tended in time to have a stultifying effect on development, so that cautiousness and conservatism grew to be characteristic.

On the other hand, a new form of economic development less fettered by tradition was being created outside the main zone of economic activity by wealthy landowners and prosperous peasants whose affluence was based on increasing rural commercialism, and by certain rich townsmen. By the mid-1840s, for example, small plants were established in the countryside for the production of such staple food items as fermented bean paste (*miso*), and *sake*. Another manifestation of this period is seen in the localisation of textile manufacture in present-day Gumma prefecture, in the northwestern reaches of the Kanto Plain. Reischauer notes that although the provincial businessmen operated at small scale, 'as a group, they probably contributed more to the subsequent modernisation of the Japanese economy than did the wealthy but more cautious big merchants in the cities'.[11]

### Summary

Industry of comparatively large scale, at least sufficiently large to have supported a sizeable foreign trade between Japan and other countries in the western perimeter of monsoon Asia, was established long before the Edo period.

5.4 Retired bus turned into a noodle stand (restaurant). Kanagawa prefecture, August 1980

# Summary

Following the early development of indigenous handicrafts which were gradually blended with influences from China and Korea, the Japanese entered upon a long period of trade and piracy. As this activity grew, imports were dominated by copper 'cash', while exports included certain raw materials and many manufactured items for which the Japanese came to earn a high reputation. Among the foremost of these were swords, and by the fifteenth century the trade in military hardware and other products had grown to significant proportions.

Such early manufacturing gave the Japanese great technical skill as well as rich experience in commercial organisation. Between the thirteenth and the early seventeenth centuries, foreign trade, both normal and 'illicit', was sponsored by some wealthy Buddhist monasteries, later by certain *daimyō* and, as merchant organisations increased in power and influence, by successful business people and the many associations into which they eventually became organised. Trade with China was of great importance to Japan but the situation was complicated by Chinese tradition, which was marked by condescension toward all trade partners. As the Japanese were not always willing or able to entertain such a relationship, exchanges between the two cultures were often unofficial. These were none the less large and important, but because of the circumstances, are difficult to document and are therefore historically obscure.

After the Tokugawa triumph in 1600, the policies of the previous administrations were continued until the shōgunate was convinced that further freedom of association between Japan and the outside world would threaten its powers of decision-making. However, as the most serious curtailment of general internal activity did not occur until after about 1615, and as anti-foreign and anti-Christian edicts were not imposed until the 1630s, there was a relatively protracted period of economic and other expansion. As agriculture was improved the population rose, cities grew in number and importance, and there was a marked increase especially in internal trade and commerce. Moreover, although the restrictions that came later were remarkably effective in sealing Japan off from the outside, internal developments, uneven because of the fluctuating powers of the regime and the uncertain nature of the economy, continued in the direction of a unified state. There were particular developments in the organisation of trade and industry whose associations often came to exert a restraining influence on the shogunal powers. In a negative way, the heavy official emphasis on agriculture, from which the bulk of the nation's revenue was derived, benefited the commercial and industrial sectors, despite eventual attempts by local *daimyō* to use them for monetary support.

Noteworthy in Edo times was the formation of large family corporations, not only as embryonic *zaibatsu* concerned mainly with wholesale operations, but also as the forerunners of retail empires, some of which are still important. Always, however, the regime and its adherents attempted to regulate commercial life. Hence, development was uneven and the

*Historical aspects of the commercial landscape*

forces of change, although finally triumphant, were never wholly effective in bringing about a true industrial revolution, or even in producing a firm background for the sweeping changes that occurred after 1868.

Industry, addressed chiefly to the home market, expanded in scale while production of many articles came to be localised by the occurrence of raw materials, the availability of labour, by concentrations of industrial and commercial technicians, or by combinations of such factors. There tended, as in the present, to be an overbalance of commercial activity in what is now the urban core region, which had certain advantages over the provinces. For the most part, this region was administered directly by the Edo Government and had a somewhat better balance of civilians to military personnel. There was thus greater freedom from official supervision and, conversely, more opportunity for commercial expression. The locus of commercial development, however, tended in time to shift into the provinces, where certain industrial specialisations came to be encouraged while at the same time such activity in the old core area was being increasingly limited by repressive policies. As a consequence, provincial business activity in the late Edo period had a specially important influence on the economic modernisation that followed the Meiji Restoration.

# 6
# Landscapes of commerce and industry, 1868-1945

The rise of Japan after 1868 to a position of world prominence in commerce, industry and military might did not come about overnight. It is nevertheless true that compared to the tempo of development in the West, the base which the Japanese had managed to establish in the 1850s, as sketched in Chapter 5, was so modest that even the most minor advances assumed an air of the spectacular, specially when it was not these aspects of life alone that were being consciously revamped; drastic changes were occurring in all directions. What is more, the pace was necessarily accelerated by external political factors over which the Japanese had no control.

The gap was not so wide perhaps in years, for the nations of the West had been experiencing important advances in industry, transportation, trade, medicine and other areas only since the first decades of the nineteenth century. But a handicap of merely fifty years was an enormous hurdle to the Japanese because of the furious rate of development occurring generally in the world as fresh approaches and ideas emanating from northwestern Europe and the United States were disseminated. In any case, the social developments that accompanied the Industrial Revolution in the Western world had been brewing for a century or more; during most of this time, despite such manifestations of 'incipient capitalism' as were noted in the foregoing chapter, Japan had remained largely feudal. Even the transfer of the governmental reins following the Restoration in 1868 could hardly be called egalitarian. The whole Meiji period movement was authoritarian in the extreme, and it was not until the early 1920s that the roots of liberal democracy, feeble and ephemeral though they may have been at the time, were truly established. This is not to belittle the agencies of change after the passing of Tokugawa power. Indeed, it is highly unlikely that the culture could have been so thoroughly transformed without them.

Aside from the matter of the international fortunes of the day, which may have been peculiarly favourable to Japan, the course of modernisation might well have been bungled had it been directed by men of lesser talents and imagination than those of the relatively small group of young patriots who assumed the leadership following the collapse of the Edo Government in 1868.[1] The extraordinary foresight, intelligence and dedication of this body of mainly moderately-ranked *samurai*, who later attained virtual immortality as the 'Meiji Oligarchs', was responsible in no small measure for the many successes registered by the new state throughout the long Meiji era and well into the succeeding Taisho and early Shōwa periods.

## Industrial developments, 1850–67[2]

Modern industrial development began in the waning years of the Edo period and was accelerated by the Perry expedition which, more than anything else, brought home to the Japanese the need for new techniques of production, especially of military equipment. Early manufacturing modelled after that of the West was therefore motivated strongly by considerations of national defence and security and was centred on iron, armaments and shipbuilding. The first of these demanded new methods of smelting to produce iron of sufficient toughness for casting into cannon. The first successes were scored in 1850 at Saga, where a modern reverberatory furnace had been constructed. It is not without interest that the technology had, as usual, been derived from Dutch texts and that those equipped with such knowledge were concentrated in localities where there was enough wealth to support such studies and their practical application. Other than the shogunate in Edo, only a few of the larger *han* had the resources for this; and of these, only Mito in present-day Ibaraki prefecture, northeast of Tokyo, was located far from the seats of learning. The other fiefs important in this regard — Satsuma of southern Kyushu, Choshu of southwestern Honshu, Hizen of northwestern Kyushu, and Tosa of southern Shikoku — were all within relatively easy access of Nagasaki.

Throughout the 1850s, the building of other reverberatory furnaces and the casting of cannon went on by fits and starts in several parts of Japan, notably in the above-named *han*. The building of modern iron furnaces provided the basic material for weapons, and during the years between the mid-1850s and the Restoration, the armaments industry grew to significant proportions, with the main centres of production at Kagoshima, seat of Satsuma *han*, as well as at Saga, capital of Hizen, and at Mito. Experiments in furnace-building and armaments manufacture took place in many other locations but, probably because of insufficient capital, few were successful. On the other hand, despite the long distances separating the major initiators, there was surprising cooperation among them in the exchange of technical information and even in financing.[3]

The shōgunate did not manage to build a functional reverberatory furnace until 1858, after which it quickly turned to the building of ships — now a legitimate undertaking, thanks to the removal in 1853 of the 218-year-old ordinance prohibiting the construction of seagoing vessels. The coming of the Perry expedition that same year had alerted the Japanese to the threat of encroachments from the West, and the Tokugawa Government soon became anxious to create a modern navy.

Shipbuilding, therefore, was the third of the new industries to be started well before the Restoration and to be inspired by foreigners or by the

6.1 Detail of small family housing on the outskirts of Sendai, August 1980

menace of foreign power. The first launchings of new-style vessels came in the period 1855–57, and the earliest ships were apparently built without the kinds of direct foreign assistance which became so characteristic of the Meiji period.

Industrial development employing, as in later years, foreign technicians and equipment, began in 1855 at Nagasaki with a naval training programme under Dutch supervision. By the early 1860s this had grown into a sizeable iron foundry and ship-repair facility. Inevitably, however, it proved inadequate to the growing demand and there followed, under French direction, the construction of iron foundries at Yokohama and at nearby Yokosuka, the latter in conjunction with a shipyard that was not completed until after the Restoration. Both Yokohama and Yokosuka assumed new functions at this time and later rose to considerable prominence in the urban hierarchy. The former quickly became the great outport city for Tokyo and the Kanto Plain, and the latter evolved as the nation's largest naval base.

Both before and after 1867, shipbuilding was far more decentralised than either smelting or armaments manufacturing, although it tended to concentrate in western Japan and around the Setonaikai. Shipyards mushroomed broadly, becoming established even at such small coastal cities along the Sea of Japan as Nanao in modern Ishikawa prefecture, or at Tsugaru in extreme northern Honshu, close to the present city of Aomori. Ships were also purchased abroad, and by the time of the Restoration there was a total of 138 Western-style craft, only forty-four of which

## Landscapes of commerce and industry, 1868–1945

6.2 *a.b.* Types of houses in upper-middle-class residential suburb of Nishi Kamakura, August 1980, near Yokohama

belonged to the shogunate. Most of these were sailing vessels. The only locally built steamships originated in Satsuma, Mito and Saga, centres of those *han* which emphasised shipbuilding, where techniques were based mainly on imported books rather than on knowledge acquired directly from foreign advisers.

## Industrial developments, 1850—67

To finance their adventures in smelting and casting, armaments-making, shipbuilding and other strategic industries, some *han* turned to commercial industrial development; even in the late Edo period factories were established where new production methods were applied and where more advanced techniques were used in the extraction of raw materials. Unlike the strategic industries, furthermore, this kind of activity depended heavily on foreign (chiefly British) assistance. For example, a modern, large-scale, steam-powered textile mill was opened in 1868 at Kagoshima, while Saga in the same year entered into a joint venture with a British concern to exploit local coal deposits, both enterprises being undertaken with the aid of English specialists, equipment and capital.

These, then, were developments typical of late Edo times. If they were inadequate to the massive expansion that took place in the following decades, it must be insisted that they were not without importance in so far as they constituted the roots of a broadbased industrial system. In brief, the Meiji Government was not forced to begin from scratch. On the contrary, it was fortunate in inheriting a functioning industrial establishment together with a corps of trained and experienced personnel, especially in the fields of modern iron smelting and fabrication, mechanised spinning, coal mining, shipbuilding and repair, and even in the operations of a modest merchant marine.

Most important, however, the pre-Restoration experiments, especially those of the shogunate itself, helped to set the stage for later modes of development. Experience in Government ownership and management, for example, was useful in formulating later procedures for creating industries and then for nurturing them to the point of viability. The same can be said of the practice of utilising foreign engineers and other technicians, or of setting up training programmes both in Japan and abroad, to provide additional qualified personnel.

Even at the time, however, these assets could hardly be compared with those of the technologically advanced nations of the West, and there is little question but that Japan entered its modern age with marked social and economic disadvantages. There were, for example, several strongly negative features in Japan's readiness for industrial expansion. There was little capital, nor was there much likelihood of any very rapid strengthening of the financial resource base, despite the efforts of a few enlightened *daimyō* to promote industry and commerce. And although the Japanese showed a brilliant perceptivity in avoiding the loss of economic controls which their Chinese neighbours had sustained at the hands of foreigners, they too were forced for a time to endure unequal treaties which prohibited the use of protective tariffs. This circumstance could only have retarded even further the development of vigour in certain industries. And although the Japanese strived to overcome this affront to their sovereignty, full tariff autonomy was not realised until 1911, at the very end of the Meiji period.

## The modernisation of the industrial landscape

Like most of the major cultural transformations deliberately induced by the Japanese, the evolution of a modern industrial landscape was brought about by the overlay of foreign or semi-foreign influence on a deep base of native tradition. The dominance of imported over indigenous elements is, moreover, very recent: many scholars, in fact, dispute that this has yet occurred. In some parts of the country traditional enterprise remains important, and from the very onset of the Meiji period, both modern and traditional industry have tended to exist side by side.[4]

Unlike Europe after the Industrial Revolution, Japan did not suffer the almost complete destruction and replacement of its political, economic and social institutions. In 1868 there was little direct involvement of the masses in the sudden passage of the Government from feudalism to an absolute monarchy with slowly emerging suggestions of democracy. As has been suggested, the propelling force in the Restoration movement was provided by a new elite composed for the most part of generally youthful *samurai*, predominantly from western Japan, and linked with certain nobles of the Imperial court, a coalition which reformed the political system but which, in retaining the old social order, used it to further the processes of modernisation. For example, as in the feudal period, the farmers were made to furnish, through taxes, the bulk of the nation's revenue, and in consequence, since some of the traditional safeguards against complete impoverishment and dispossession were no longer effective, their economic circumstances worsened, as can be seen in Chapter 2.

On the other hand, this system supplied the Government with needed capital which, in general, was then honestly and scrupulously administered. Modern industrial technology and equipment were purchased on the world market from the most reliable sources of the period. Foreign advisers were invited to Japan at high salaries to supervise initial development, but no time was lost in training an indigenous work force in order that the tenure of the expensive visitors would be minimal.

Two factors in particular helped to improve the prospects of industrial modernisation.[5] One was the existence of a large body of potential factory workers who were willing to accept low wages. After the Restoration, population growth accelerated (Table 1.3), and though it was several decades before pressures became severe, there were always surpluses in rural areas. Furthermore, if by 1868 ordinary citizens had not been educated at least to the point of basic literacy, steps were taken soon after to remedy the condition. Initial plans for some kind of mass schooling were laid as early as 1873, and in the 1880s there was a shift in emphasis to general elementary education for all, a movement that was boosted in 1890 by the famous *Imperial Rescript on Education*, a brief document that, from that time forward until the end of the Second World War, every school child was required to memorise. By 1905, therefore, the Japanese were able to make the claim, highly unusual at the time, that 95 per cent of its children of statutory school age were receiving an elementary education.[6] The second factor of great importance in early Meiji industrial development was the willingness of the public to follow the Government's lead and accept new ways of life. To be sure, adaptability, particularly on the part of the intellectuals, is a nearly constant theme in Japanese history, but the era of the Meiji Restoration and its aftermath is surely one of the supreme examples of this remarkable national characteristic.

Compared to what was going on in most parts of the West at this time, however, Japanese industrial growth in the late nineteenth century was overbalanced in favour of military production. The slogan of the new Meiji elite, despite vigorous efforts in the direction of education and general public enlightenment, was *fukoku kyōhei* ('national wealth and military strength'). Reflecting this sentiment, individual living standards, especially in the countryside which contained the bulk of the citizenry, were either neglected or depressed, while the growth of a domestic market was kept under tight control. Many major reforms were not undertaken until after 1945.

The stages of economic expansion since the Restoration have been delineated by many. One source views the time between 1868 and the present as consisting of six segments, beginning with a stage of *infancy*, from 1868 to 1894, the first twenty-six years of the new regime.[7] During this period, as stated, the Japanese emphasised modern defence and at the same time addressed themselves to the then difficult task of shielding their new and often feeble industries against foreign economic invasion. It was at this time that the Government took a particularly strong hand in projects that required unusually heavy capital investment. The whole gigantic scheme to open, settle and exploit the riches of Hokkaido is a case in point. Although the first ventures into the northern island had come very early, by the nineteenth century permanent settlement was still confined to isolated coastal areas and to the extreme southern districts, with the main thrust, including the founding of Sapporo and other important inland settlements, coming from Tokyo following the Restoration.

But the expenditure of effort on the Hokkaido settlement operations in the 1870s, on top of widespread Government-sponsored industrial development in Old Japan, and the suppression of a rebellion in 1877, placed a severe strain on national finances. A period of retrenchment consequently intervened in the early 1880s. Industries, except for strategic industry, communications and public services, were slowly sold to private corporations, often to the forerunners of the *zaibatsu*, at token prices. This had the salutary effect of stabilising the national economy, and was followed by an upsurge of industrial activity, at least in certain fields. For example, the textile industry gained much strength at this time, and there was a rash of activity as large spinning companies were created, modernised and enlarged. The boom in textiles was followed by the establishment and quick growth of many other industrial enterprises in diverse fields, with private single undertakings soon tending to become absorbed by huge corporations — a proclivity which, according to Reischauer, developed with unusual rapidity compared to what had happened earlier in the West.[8]

The second stage, that of *initial growth*, extended for twenty-four years, from 1895 to 1918, and was marked by the rise to considerable autonomy on the international scene on the part of the textile industry. Concurrently, heavy industry was given its start, in the sense of modern, large-scale operations, with the completion in 1901 of the Yawata Iron Works, planned and built by the Government on the latest technological principles and forming a nucleus in northern Kyushu for the nation's premier heavy industrial complex. As this period began, furthermore, Japan was successfully bringing to a conclusion the first of its modern wars, for which, according to the rules of the day, there were substantial rewards. The Sino-Japanese War of 1894—95 resulted not only in elevated prestige for the nation and in the acquisition of foreign territory, but also the Chinese were made to pay indemnities that more than covered the costs of the venture, thereby allowing the Japanese further latitude for industrial expansion. As a result, most of the remaining state-protected industries were able to break their ties with the Government. By the end of the nineteenth century additional advances were reflected in the ending of the need to import foreign yarn and even in the initial invasion of Japanese textiles into the Chinese market. This period also saw significant industrial growth in other fields such as sugar refining, fertiliser production, gas and electricity generation, and in transportation, especially in railroad construction and consolidation. Meanwhile, the Government-owned and operated armaments industries were also progressing.

It must be noted, however, that these advances, while momentous at home, were relatively small on a world scale. Japan had not yet begun to play the role of a serious competitor to the West in international trade, although its first important opportunity was close at hand.

Meantime, the Russo-Japanese War of 1904—5 and the annexation of Korea in 1910 tentatively eliminated certain competition in East Asia and helped to provide other benefits. In this case, however, the costs were also high and the economic gains proportionately less, although the conflict did result in further economic stimulation, particularly of the armaments industry, of marine transport and of industrial technology in general. The development of financial institutions was also accelerated and was an important step in this stage of economic evolution.[9]

The first stimulus to industrial advance that was important by world standards was the First World War, which saw Japan, as one of the Allied Powers, exploit the preoccupations of its Western rivals with the situation in Europe, to become a major supplier of cotton cloth, munitions and other material and nearly to double its merchant marine tonnage.[10] As a consequence, by the end of 1919 Japan had transformed itself from a debtor to a creditor nation, and the economy had truly come of age.

During the long initial phase of industrial evolution which covered a timespan of fifty years, as outlined in the two sections above, textiles was the leading industry and during this half-century there was much change in its location. As operations became mechanised, the traditional silk industry, which had been largely characterised by handwork, was particularly affected. In the early and mid-Meiji years, silk production had become geographically very diffuse, although there were concentrations of activity in northern Kanto and in southern Tohoku. But as the industry was modernised there was a shift in location to central Honshu, especially to Nagano prefecture. Moreover, as the scale of the industry was enlarged, certain towns in the new locations grew into important manufacturing cities.[11]

After 1890 the cotton industry also became a large-scale private enterprise of great significance, with modern mills locating first in central and western Honshu, particularly in Aichi and Osaka prefectures, where there was early specialisation in the production of raw cotton. Gradually, however, as imported cotton supplanted the homegrown product, the factories tended to migrate towards the great international ports, especially to those that were oriented toward China. The growth in importance in the cotton industry of Osaka, Kobe and Nagoya exemplifies this trend.

Heavy industrial location was also affected as this period progressed. In addition to the northern Kyushu centres around Yawata, iron and steel production became characteristic of Muroran in southern Hokkaido and of Kamaishi on the Pacific coast of Iwate Prefecture in Tohoku. The main factor in all these coastal locations was the presence of raw materials, chiefly coal in Kyushu and Hokkaido and iron ore in modest amounts at Kamaishi.

As industry matured and its role as a landscape feature became more prominent, many cities, even in the outlands, came to exhibit this function, and as the larger companies came to control the most modern, labour-intensive enterprises of the day, a corporation-controlled form of urbanisation spread widely, particularly into areas of mineral wealth. Cities such as Omuta of northern Kyushu, Niihama of northern Shikoku, Nobeoka of southeastern Kyushu, and Nagasaki, became in marked degree company towns dominated by the great *zaibatsu*, of which Mitsui, Mitsubishi and Sumitomo were the leaders.

Stage three, that of rapid growth and first prosperity, the ten-year period from 1919 to 1929, was conspicuous for the strengthening of Japan's competitive position in spinning, both through the importation of the most advanced foreign equipment and technology and by means of the first Japanese industrial innovation to attract major attention abroad. The patenting of the Toyoda automatic loom in the 1920s was one of the earliest indications of Japanese inventive genius and technical ascendancy in a worldwide context. Proof of this ascendancy came in 1929 with the purchase by an English company of rights to manufacture the Toyoda loom, and this was followed by the dispatch of a team of Japanese experts to supervise the construction of the device. This was a startling realisation, as the methods employed by the Japanese to hurriedly industrialise had led to their being viewed internationally as a nation of unimaginative

6.3 Mitsubishi shipyards, Nagasaki, showing vessels under construction

copyists — a stigma that was not to be finally lived down until the 1960s.

During this period advances were made in many other industrial areas. Steel manufacturing and the machinery industries were favoured by Government contracts for ships and other heavy products. The automotive industry gained importance and the aircraft industry was launched, while there were similar increases in the output of railroad, power-generating and other heavy equipment by industries that had been under development, often with Government protection, since the nineteenth century. Likewise, the heavy chemical industry achieved the capability of supplying on a large scale agricultural chemicals such as aluminium sulphate to help in the advancement of the agricultural sector.

In the 1920s the exportation of raw silk and of manufactured silk products, especially of hosiery to the United States, was of particular importance. This circumstance was especially related to two factors. In the first place, after 1925 economic depression deepened in rural Japan and farmers became more eager to supplement their incomes through sericulture. In the second, the silk industry had long been the very backbone of the export trade. It had, in fact, been the leading revenue-earning export since the onset of Japan's industrial modernisation. Lockwood has estimated that between 1870 and 1930 the raw silk trade alone was able to finance as much as 40 per cent of Japan's entire imports of foreign machinery and raw materials.[12]

During the 1920s Japanese industry had begun to catch up with that of the West, only to founder in the wake of the stock market collapse of 1929 and in the subsequent stagnation of international trade. This sequence of events betokened the fourth and last period before the Second

World War, which encompassed the fifteen years between 1930 and 1945 and was generally one of war-based enterprise. Being particularly vulnerable and feeling the effects of these reversals perhaps sooner and more deeply than many other nations, Japanese turned at home to the expedient of industrial rationalisation, a movement of which the silk industry, including its agricultural foundations, was a particular target. In 1930 a Bureau of Industrial Rationalisation was established by the Government, ostensibly to aid depression-ridden industries but with wide regulatory powers designed to increase efficiency and lower costs. Although its long-range objectives were not met, there were vast improvements, albeit at the expense particularly of many smallscale undertakings. In the following year, the passage of the Major Industries Control Law, in effect, enabled the Minister of Commerce, in certain cases, to restrict competition in large-scale industry, a measure which had only limited results.[13]

In any case, the need for such measures was lessened after 1931 by an industrial resurgence occasioned in part by a massive devaluation of the yen at the start of this period, which lowered the whole economic structure and allowed Japanese manufacturers, through reduced operating costs, to undersell on the world market. This comparative advantage was, of course, much to the consternation of other industrial nations, whose factories all too often had grown idle. These circumstances provoked outcries of frustration and increasing bitterness on the part of Japan's competitors and, at home, social, economic and political unrest. In 1931 the Japanese Kwantung Army advanced into Manchuria to launch nearly a decade of tumultuous military adventures by the so-called totalitarian powers, and soon a vital part of China had become a Japanese-controlled vassal state. In Japan, the Democratic Party Government was gradually overwhelmed by an undisguised military oligarchy, while economic orthodoxy was replaced by authoritarian controls. By 1936, reactionary military elements in Japan had managed to foment a *coup d'état* which, although largely unsuccessful, set the stage for full-scale war with China beginning in the summer of 1937. From that time until the end of the Second World War in 1945, the nation was on a complete war footing, with militarism as the characteristic political mode.

The economic stimulation of the early 1930s, regardless of its untoward repercussions abroad, had important beneficial effects on the development, and particularly on the diversification of industry. In the early years of the decade, when normal peacetime production was intensified to satisfy a larger export market, textiles were the leading commodity; during this period, incidentally, the silk trade was overshadowed in terms of financial rewards by cotton, rayon and wool. Industrial diversification in the years from 1930 to 1937, meanwhile, was reflected in sharp increases in the exportation of an immense variety of predominantly consumer goods. It was on account of this latter trade that the Japanese, usually at the instigation of foreign buyers, earned their reputation for shoddy goods, a reputation which was especially strong in the United States. On the other hand, increases in capital construction in Manchuria during these years helped to raise the export prominence of industries whose products had not hitherto been seen abroad. These included the output of the metals and machinery industries and even such things as vehicles and scientific instruments.

These and related developments reflected a major change in the direction of the export trade. Whereas in the 1920s the United States had been Japan's major customer, after 1931 and the rise of industries based on the nation's new expansionist policies there was a decided shift in the destinations of Japanese products. By 1936, largely because of the depression which eliminated the demand for silk, the market for Japanese exports in the United States had shrunk to nearly half that of a decade earlier. This was balanced, however, by the expansion of the war-generated home economy and by a transfer of Japan's overseas markets to other areas, notably northwestern Europe and such Asian nations as India and China (especially as Manchuria was now under Japanese control), with sizeable increases in trade with Africa, South America and Oceania. There was, in short, a general displacement in export dependence away from the United States and the West in general, toward Asia.[14]

Imports, meanwhile, underwent large increases to accompany the development of the export trade. In addition to increasing numbers and quantities of raw materials, these included such finished goods as machinery and vehicles, inasmuch as Japanese technological advances, significant as they were, were unable to keep pace with rising demands. At the same time, serious trade imbalances developed out of the circumstance that the chief source of their imports continued to be the United States, notwithstanding the fact that the latter was no longer Japan's most important customer. These imbalances in trade were to become, of course, prime causes of the Pacific War.

Between 1930 and 1936, with variations according to the type of industrial activity, economic conditions in Japan generally tended to improve or to remain at least stable, specially in the regions most accessible to the now more rapidly growing urban core. There was little sweeping prosperity for the average citizen but mostly there was no such economic decline as was occurring in much of the outside world. Only the agricultural sector remained in a depressed state, with chronic conditions of overpopulation and underemployment, but job opportunities in the cities provided at least some outlet. In the main, therefore, the deepest depression years for the world at large were ones of growth for the Japanese. And this, Lockwood asserts, was no mean achievement for the time.[15]

The trends of industrial growth established in the first half of the 1930s, for the most part, continued in the second except that after 1937 these became increasingly influenced by two factors: Japanese military expansion in China and eastern Asia, and rising resistance in the West — in the United States particularly — to the purchase of Japanese manufactured goods. As the terms of trade gradually worsened for Japan, and as feelings of hostility towards Britain and the United States mounted, industry at home turned more and more to the production of military equipment and supplies. And although there was continued urban growth and some accompanying prosperity, the directions of industrial evolution were constantly away from peacetime development.

By 1940 the Japanese had become heavily committed to an industrial empire based on war. In that year, 17 per cent of the entire national output was for war purposes. By then the die was cast for a showdown which, in the years 1944 and 1945, saw the end of a long period of formation and growth in which there had been accomplishments that often were no less than miraculous, considering the brevity of the experience and the dearth of resources. When full-scale war finally came, and as the world soon learned, Japan had unquestionably attained major status as an industrial superstate. It was the only non-Western, self-directed nation to have achieved this distinction.

Industrialisation, by this time, had also resulted in many fundamental changes in the Japanese way of life. Traditional Confucianist patterns of interpersonal behaviour remained the norm, but these were being undermined by such factors as the rapid growth of the urban population, a new social mobility, and crowded if not unstable living conditions. All these changes, singly or in combination, favoured the abandonment of tradition and resulted in a considerable loosening of the fabric of society. Cities were also providing an unprecedented anonymity for the individual. Privacy of sorts, always a rare commodity in Japanese culture, was becoming known and appreciated as never before. To counter such influences, the Government after 1937 instituted or revived various systems of social control. Farm communities had long been accustomed to a tightly regimented cooperative way of life, and now these techniques were applied also to the cities. There were community neighbourhood associations (*tonari-gumi*), organised usually by wards or by small groups of family dwellings, whose operations were designed to implement Government directives and services and to see to the orderly conduct of public and even private affairs. These included such matters as health, family and social behaviour, and even personal conduct and thought. In addition, to guard the national security, there was a pervasive network of ordinary and special police who were able to perform quick and efficient personal surveillance. On the positive side, during the war and especially after 1944, when bombings and disruptions were of daily occurrence and urban populations were evacuated into the countryside, it was the neighbourhood associations that provided services that helped to maintain some semblance of normalcy. Even in the bleak postwar years between 1945 and 1947 these organisations continued to function in various ways, notably in the dissemination of information and in the distribution of the meagre and often bizarre rations of food and other supplies.

The interwar period, from 1918 until 1945, saw the completion of the basic contemporary industrial structure.[16] Distinctive in this period were developments in electrochemicals, chemical fibres and textiles, and in electrical machinery, for all of which the chief promoters were *zaibatsu*, either those of historical prominence such as Mitsui and Sumitomo, or those more newly developed, such as the Hitachi corporation, that had been particularly spurred by a wartime economy.

During this period further areal changes in industry occurred. Chief among these was the expansion of industry into the interstices of the major urban districts, a development which accompanied the rise of such new industrial nuclei as Kawasaki and Tsurumi in the Kanto area and Amagasaki and Nishinomiya in the Kansai. In these and other centres, frequently on reclaimed land with man-made deepwater access, were built iron and steel mills, electrical machinery and chemical manufacturing plants, and food, beverage and other processing facilities.

A concomitant trend during this twenty-seven-year period was a movement of industry into the countryside, largely in response to such primary activity as coal and copper mining, limestone quarrying and hydroelectric power generation. As some of these depended heavily on cheap sources of power, the locations were often in such remote areas as central Honshu or southern Kyushu.

In consequence of these developments, stretches of farmland or of traditional settlement between Tokyo and Yokohama and between Osaka and Kobe, were turned into diversified industrial districts, to which, especially in the latter case, were added the functions of dormitory settlement for commuters to the larger cities. The atmosphere in such places was thus much altered. The bayside areas south of Tokyo and Osaka respectively became largely industrial with apartment enclaves for employees, while inland, and in marked contrast, settlement was changed from traditional agrarian to modern suburban as interurban rail lines were constructed. The site characteristics of Kansai for dormitory settlement, which included a variegated landscape backed in the Kobe area by the steep walls of Mount Rokko, provided a setting for some of the most elegant suburban housing of the prewar period; while in the Kanto area the more beautiful suburbs tended to be within the Tokyo city limits, with some inclination towards expansion to the west along the Chuo line of the Japanese National Railways. South of Tokyo and inland, except along the main electric rail lines, a traditional agricultural style of life persisted until the 1950s.

It was also during this period that company towns in more remote locations, such as Nobeoka of Kyushu and Niihama of Shikoku, grew to be more heavily industrialised and that this class of towns became augmented by other primary and secondary industrial centres of similar background but sponsored by 'newer' *zaibatsu*. Of these last, the city of Hitachi, northeast of Mito, is the foremost example. Such cities were unusual in that the control of commercial life by *zaibatsu* tended to inhibit the growth of tertiary industry. Hence such centres, known to some as 'blue-collar' towns, were generally without the kinds of elaborate shopping facilities that usually marked other cities of comparable size.

6.4 Main street leading from railway station, Sendai, August 1980. This is typical of a central urban area which was devastated by fire-bombing in 1945 and rebuilt. Buildings are new, streets are wide with broad sidewalks, old tram lines have been removed

## Some effects of the war on the industrial landscape

The urban-industrial landscape was, of course, drastically rearranged by war. Most cities of note, as has been mentioned earlier, were subjected to 'saturation' bombing, usually in the form of one-night incendiary raids, while the larger and especially the industrial centres suffered repeated attacks. As Japanese traditional structures were often of wood and other inflammable materials, there was grievous destruction of material as well as loss of human life. The prefectural capitals, being generally the more vital regional centres, were the principal targets of these onslaughts, and while Hiroshima and Nagasaki drew most of the world's attention for their destruction by nuclear arms most of the main cities were razed almost as completely. Since the built-up parts of cities were then much more compact than those of the sprawling communities of the present, the proportions of total areas destroyed were very high. A few major places were not bombed, but the exceptions were usually those which lacked significant industry, particularly those which had important historical background — attributes that, most conspicuously in the case of Kyoto, were brought to the attention of the Allied military command by a small but influential cadre of Japan specialists employed as advisers in Washington. Kanazawa, Kurashiki, Nara, Matsue, Takaoka and Takayama are included in the same category. Niigata, despite a large amount of heavy industry and important port functions, was selected as a target but was spared because of cloudy conditions on the night of the raid, followed by an abrupt end to the war. Hokkaido cities survived for similar reasons and also because they were far from the urban mainstream. On the whole, however, the industrial establishment was as much affected by the severing of external supply lines as by direct destruction by Allied bombers. Total production figures for the war years show massive disruptions in the normal output and flow of vital commodities, as well as the virtual cessation of manufacturing for civilian consumption. Reischauer asserts that in the last days of the war the average Japanese was consuming less than 1 500 calories a day.[17]

The early war years therefore represent the culmination of a long period of industrialisation, which was temporarily halted by the physical destruction of the final war years and by the five or so stagnant years after the surrender. Most important industrial operations, for reasons to be adduced in the following chapter, were not resumed until at least the late 1940s, and by the time of initial recovery in the 1950s much of the prewar industrial facilities had been replaced or augmented. There was, in any case, a distinct end to one phase of development and a long pause before activity was resumed. Consequently it appears fitting to devote a separate chapter to the restructuring of the industrial domain and to the evolution of the contemporary industrial landscape. Phases five and six in the *Japan Times*'s history of industrial growth — from the Occupation to the end of initial reconstruction, and from 1956 to the present, respectively — will therefore be presented in Chapter 7.

## Summary

The initial transformation of Japan from a premodern to a modern industrial state encompassed a period of some ninety years and when this first phase ended in the mid-1940s there were, understandably, many changes — social reforms in particular — yet to be accomplished. None the less, the achievements, brought about by an extraordinary combination of good fortune, brilliant leadership and public devotion reinforced by proud tradition, were remarkable in the extreme.

Some of the foundations of modern industry were established in the 1850s by the more progressive *daimyō* and by the feudal Government in Edo. It is worth noting that in many cases these beginnings were made with the help of textbook knowledge alone. Inasmuch as modern armaments for national defence were the main goal of these efforts, initial successes were in the areas of iron production and weapons manufacture, followed somewhat later by development of ship construction and repair.

The recruitment of foreign advisers and the importation of key equipment from abroad likewise began in the 1850s, becoming a standard practice in the early decades of the Meiji period. Even before 1868, moreover, local *daimyō* were upgrading commercial industry such as textiles by employing foreigners as technicians and initial managers in an effort to pay for the establishment of the more capital-intensive defence industries.

When Japan rejoined the world in 1868, however, the nation was well behind the West in most aspects of modernisation. There was little capital, a dearth of technical knowledge, and only the most modest accomplishments; in addition, there was the handicap of Western domination through unequal treaties. Modernisation, on the other hand, was much favoured by the traditional social structure which encouraged patience on the part of the public, whose living standards were only marginally improved, yet who were being made to provide the bulk of the national revenue. The leaders of the new state, exemplary themselves, were careful to extol such traditional cultural values as frugality, honesty and subservience to authority, all of which were invaluable assets during the whole process of industrial modernisation. By these and other means the work force, which was gradually improving in basic education, remained tractable while at the same time it was being asked to endure the postponement, in the interests of national strength and security, of many social reforms.

Six stages of economic development can be recognised, the first being one of infancy, in which industry was nurtured by the Government into viability and then transferred to private interests. Industries gaining their first momentum in this twenty-six-year period included the strategic industries, communications and other public services and especially textiles, in which the trend towards the consolidation of companies into giant *zaibatsu*-style corporations was well established by the 1890s.

Stage two, of initial growth, saw the establishment of more modern heavy industry, the continued growth of textiles and many such businesses concerned with mass production, all of which went to elevate the nation's position as an industrial power. During this twenty-four-year span, wars contributed much to Japan's industrial progress, beginning with the Sino-Japanese War and following with the Russo-Japanese War, the annexation of Korea and ending with the First World War. Without being an important combatant but aligned with the victors, Japan profited by increased industrial sales to the belligerents and by 1919 emerged as a creditor nation with substantial and burgeoning industrial power, even by Western standards.

Areal changes in industrial location from the onset of modernisation to 1918 included a concentration of the silk industry in central Honshu and the growth of manufacturing towns based on this activity; a movement of the cotton industry towards port cities which served especially the China trade; the development around raw material deposits of new heavy industrial centres in Hokkaido and northern Honshu; a general intensification of industry as a major landscape feature, and the growth in remote places of *zaibatsu*-controlled heavy industrial cities.

Stage three, of rapid growth and first prosperity, featured the rise of the spinning industry to international prominence. This was marked by the first successful Japanese technological breakthrough in the patenting and sale abroad of highly sophisticated spinning machinery. Many other industries, including motor cars, aircraft and chemicals, grew as well, but the leader in terms of revenue, particularly from the United States, continued to be the silk industry, which was able to support the purchase of many vital imports.

Stage four, the fifteen years before the end of the Second World War, saw Japanese industry, reacting to the depression which, in particular, reduced the silk trade with the United States, turn gradually back to strategic production to support widening military operations in East Asia. The early 1930s were a time of yen devaluation, followed by industrial expansion and the proliferation of trade in sundry items, often cheaply made, which damaged the image of Japanese manufacturing in the eyes of the outside world. On the other hand, a shift in the location of the chief markets, from the United States to Asia in particular, helped to promote a rise in the production of durable goods for export. At the same time, dangerous imbalances of trade were developing with the United States (which continued to be Japan's leading supplier of raw materials and other goods), and this accelerated the advent of open hostilities. After 1937 the Japanese economy moved rapidly to a full war footing which increased employment and spurred urbanisation but which failed to bring fundamental improvements in living standards, especially among farmers. None the less, the 1930s, contrary to the experience in most of the world, were a time of important industrial and economic growth for Japan. As such, they were a period of remarkable achievement.

Industrialisation resulted in major changes in the Japanese life style, but the benefits of city living were limited in the late 1930s by Government programmes of social control. By the start of the war, therefore, personal freedom had been effectively eliminated, and this was followed in 1944 and 1945 by the almost complete annihilation of cities. Industrial production and the flow of vital materials were also vastly curtailed, until at the war's end the average urban survivor of the bombings was poorly clothed and fed and was often homeless. The first great era of Japanese industrial modernisation had come to a dramatic close.

The period from 1918 to 1945 saw the establishment of the modern industrial framework of the nation and the continued spread of industry throughout the land, especially between the greatest cities and their outports. These areas were transformed in two ways, the littorals becoming devoted to heavy and other types of industry as well as to rudimentary labour settlement, and the inland portions to either traditional agrarian living, interspersed with suburbanisation along developing electric rail lines, as in Kanto, or to a predominantly dormitory-style settlement with touches of suburban elegance, as in Kansai. The growth of company towns in remote places continued also, with new corporations joining the older *zaibatsu*.

# 7
# Reconstruction and the growth of the contemporary industrial landscape

The Japanese industrial landscape of today differs from its earlier forms in a number of important respects. The differences are often ones of degree, for the arrangement of the central industrial districts strongly reflects the patterns of the past, but there is more to the new scene than a mere enlargement of scale.

Industry, for one thing, has taken on a more dominant role in areas where it was formerly of little consequence as a landscape feature. The shores of the Setonaikai, for example, in Shikoku and especially in southwestern Honshu between Okayama and the Straits of Shimonoseki, are far more industrialised than before, despite the presence of important pockets of heavy industry which have produced armaments, chemicals and fertilisers, ships and the like since well before the war. The coast of Yamaguchi prefecture contained one such series of enclaves around local cement and other raw materials, while shipbuilding, together with other manufacturing, including light industry such as textiles and clothing, was common around many ports, as on the Kojima peninsula of southern Okayama, or in the area around Hiroshima and Kure. And in northern coastal Shikoku around Niihama, Imabari, Saijo, Matsuyama and elsewhere, there had long been mining, metallurgy, electrical, chemical, textile, and other industrial specialisation. Before the Second World War, however, such industry as had arisen was for the most part confined to limited areas which were interspersed among farming and fishing communities of great beauty and charm, situated as they often were in association with magnificent mountains and seascapes. But since the 1950s the old industrial areas, in addition to being modernised and re-equipped in various and often unprecedented ways, have become extended along the coasts. New sites have also arisen, frequently on decommissioned and converted salterns, or on freshly reclaimed lands in foreshore locations.

Japan's new industrial landscape is thus more extensive than before the Second World War, although its central core is essentially that of the past. At the same time there has been sizeable growth of new industry, especially in the coastal parts of Tokaido, Chugoku, Setouchi and northern Kyushu, with important and unprecedented industrial agglomerations even in Tohoku and Hokkaido. Only the coasts of the Sea of Japan, westward of the Noto peninsula, those of southern Shikoku and of northern and

# Reconstruction and the growth of the contemporary industrial landscape

7.1 Modern coal- and petroleum-powered thermal electric plant in Shiogama, the port of Sendai, August 1980

eastern Hokkaido are without significant industry. Even in some of these areas development is contemplated.

In a more fundamental way, Japan's international standing as an industrial nation has not merely improved but has undergone a swift reversal. It was only a few decades ago that Japanese industrialists were being reviled (often perhaps unfairly, since it was common for foreign buyers to encourage the practice) for flooding the world market with goods of inferior quality and blatantly unoriginal design. Now, on the other hand, their successors in postwar Japan are universally recognised as producers of commodities which are at once innovative, exquisite in conception and of proved dependability, and which run the full range from capital goods to precision articles of high prestige. Moreover, in the United States, where only a short time ago customer resistance to Japanese imports was specially strong, the purchase of expensive durable goods from Japan is commonplace. The United States continues to be Japan's most important trading partner, with terms of trade often in Japan's favour, and any major break in this pattern, regardless of temporary differences of opinion about the conduct of economic exchange, seems unlikely in the foreseeable future.

Industry, as stated, has spread inexorably into the hinterlands of most cities, creating an atmosphere of continuity, whereas formerly cities may have been separated by open, usually agricultural, land. Inland, as stated, the paving of main roads, accomplished largely in the late 1950s and early

1960s, has resulted in the proliferation of modern light industrial plants which often flank the highways and reach their heaviest concentrations as distances between nodes of settlement narrow. Some of this more recent construction, moreover, is well within convenient access of farming areas, thus providing opportunity for part-time non-farm employment. But even if distances to supplemental jobs are great, daily commuting is usually possible to the newly establishing industrial subregions or to local cities, thanks to improved roads and means of travel. On the other hand, in extreme cases of isolation, such as in parts of Iwate or Yamagata prefectures of Tohoku (northeast Honshu), farmers seeking additional income from labour in factories, offices or construction, may be forced to spend part of the year away from their families. In fact, social stresses attendant on such practice are commonly utilised as themes in popular literature, including television drama. But on the whole, commuting from farms to other kinds of work, as mentioned repeatedly in the foregoing, has become very general.

Following the lines of Chapter 6 of presenting Japanese economic and industrial progress in a series of specific periods, the years since 1945 are divisible into two periods, with the first covering the time from the surrender through the years of first reconstruction, roughly from 1945 to 1955.[1] Actually, reconstruction did not begin immediately after the cessation of hostilities. For one thing, the Japanese feared a heavy toll of capital and facilities through reparations, and as this matter hung fire for some time there was a general reluctance to begin new construction until at least 1948. For another, the immediate postwar years were ones of confusion, disillusionment, and other feelings that collectively were conducive to a spirit of public apathy; and this mood was only gradually dispelled. Meanwhile, Occupation reforms were systematically attempting to uproot such previous economic, political, religious and other cultural attributes as had led the Japanese to assume an arrogant, expansionistic posture in world affairs, and the goal was to lay the foundations of a modern and democratic force in eastern Asia. One observer has stated that there was

> not only a sincere soul-searching . . . with respect to prewar nationalism but also, in a typical pragmatic about-face, a veritable enthusiasm for anything American. The democratisation of Japanese industry — the dissolution of the *zaibatsu*, diffusion of shareholdings and the establishment of a strong labour movement — was thus carried out with much good will. Management eagerly absorbed new American techniques and methods, from modern accounting to public relations and marketing. Sooner than anyone expected, the Japanese economy gained strength, and Japanese firms began to flex their muscles in the international market.[2]

The first industries to resume significant activity were iron and steel, coal mining, ammonium sulphate production and electric power generation; these were soon followed by textiles and by certain other light industries, a surprising number of which remained intact. The loss of a military market sent producers scurrying to find new uses for old products, and the results were a boon to the commodity-hungry, abroad and in Japan. Also heralded was the resuscitation, refurbishment and, often, the rebirth of long-dormant, large-scale, mass-production capabilities, of which the sewing machine industry was a forerunner. Japanese enterprise, coupled with vigorous entrepreneurship on the part of a contingent of foreign traders (notably Americans), combined to produce very large and rapid increases in this field. By 1948, production of sewing machines had surpassed prewar levels and six years later, in 1954, the 400 000 units exported to the United States made up nearly half the total number imported that year.[3]

Moreover, even before this first phase of postwar development had ended, the Japanese had begun to produce a number of technologically advanced articles for which they have since become internationally famous. These initial successes were owing to a number of factors — in the first place, to the native perspicacity and imagination of the Japanese themselves, and in the second place, to such factors as United States aid, prodding by the Occupation authorities, and (thanks to greatly liberalised political and economic conditions) the rapid development of a home market. Many of these were virtually new departures and were based on the realisation, stemming from the late 1940s, of the comparative advantage of having a large supply of willing and highly qualified labour, easily adaptable to modern large-scale assembly-line techniques for producing a

7.2 Industry and urban sprawl extending together over the entire flat land surface in the northern margins of Hiroshima, southwestern Honshu. The large factory in the centre is the Mazda Automobile Plant

## Reconstruction and the growth of the contemporary industrial landscape

broad variety of precision articles, from cameras and electrical goods of all sorts to scientific instruments and equipment. Nearly all such enterprise had some prewar background, but the supply of products had been limited to the military, or at best to the Japanese empire market. Japanese radios, for example, were much seen in Japan in the 1930s, along with other electricalware, made either under joint agreement with foreign, usually European, companies, or by purely Japanese concerns. But there had been little of the sort of innovation that might have been stimulated by a competitive international market.

The optical goods industry had likewise had early foundations, but by the time of the Second World War the output was small and limited to military needs, or to such common items as lenses for eyeglasses. In any case, the Japanese themselves often preferred foreign, mostly German, products, and the few such articles made in Japan were heavily dependent on German technology and design. Evidence of this was seen in the years following the war when, as impoverished Japanese exchanged their capital possessions for inflated currency, the secondhand camera shops of Kyoto, Osaka, Tokyo and other cities were filled with a great assortment of equipment, mainly of German manufacture.

Meanwhile, exemplifying the growth occurring in many industries during the first postwar years of reconstruction, the photographic industry, prompted by heavy demand from the relatively affluent Occupation forces and later by a rising market in the United States, was rapidly augmented, chiefly in two centres, Tokyo and Osaka. In both places there had long been large corporations dealing in optical and photographic equipment, such as the Japan Optical Corporation (Nikon), of Tokyo, and Minolta of Osaka. In the latter 1940s and early 1950s, however, dozens of new manufacturers of cameras and related equipment appeared. Meantime, other concerns, usually of large dimensions and often with strong prewar roots, specialised in photographic chemicals, film and related products, or in mechanical components, such as shutters or electronic fittings. In the subsequent period of economic growth, when quality control became a central theme and manufacturing costs rose, many of the smaller companies, which frequently turned out excellent equipment, were absorbed by the larger. In a few exceptional cases, minor makers were able, largely through a combination of zeal and good fortune, to survive and grow in stature. The Yashica Company, for example, began as a small enterprise specialising in twin-lens reflex cameras of derivative design but of efficient and sturdy quality and low price. Manufacturing took place near the city of Okaya in Nagano Ken, in a region known previously for its spectacular alpine scenery and, as stated in the previous chapter, for light industry, especially in the cities on the northern and eastern fringes of Lake Suwa. In the 1920s this area became a centre for silk textiles, and in the 1950s, thanks to the energies and intelligence of key personnel, the Yashica company grew into an industrial giant, contributing heavily to the transformation and economic elevation of the entire region, its products becoming ever more technologically advanced, to the point where it has now become allied with the great German Zeiss optical corporation and its operations expanded to the larger industrial centres.

Opportunities for the establishment of many new such light industries that were particularly suited to Japan's immediate postwar conditions, as well as to such cultural traits as the capacity for hard work, manual dexterity and sensitivity to extreme detail, were thus presented by circumstances; so strong was such development that in the years following the war it was widely held that future industrial growth should be entirely along these lines. There was much talk of Japan as the 'Switzerland of Asia'.

However, as cold war pressures multiplied in the late 1940s, and especially after the start of the Korean War in 1950, the Occupation changed its emphasis from ideological to economic reform. It had become apparent by this time that Japan would require heavy as well as light industry to insure that it remained a key force in the 'free world's' outer defence lines. It had become equally apparent that industrialisation in the fullest sense was necessary to lift the national economy to a level commensurate with the size of the population and with Japan's global importance. The redevelopment of any long-range military capability was illegal according to the postwar constitution and hence was frowned upon by the Occupation. In any case, such would have been contrary to public opinion in Japan, which remained firmly disenchanted with militarism. On the other hand, after about 1948, as a result of the cold war, there was mounting pressure for the establishment of a 'self-defence force', in effect a small but modern army and other military services. A National Police Reserve (*Keisatsu-Yobitai*) was formed on 8 July 1950, to be followed in August 1952 by the National Security Force (*Hoantai*), and finally, on 1 July 1954, by the National Self-Defence Force (*Jieitai*). After 1950, therefore, there was open emphasis on such heavy industry as iron and steel, automotive and marine equipment, and machinery production, and since the construction of these capital facilities, many of which were new, would require a long period of time — at least the remaining years of the decade even for the initial phases — there was an unusual and protracted interval in which modern light industries were able to occupy the limelight. These became highly inventive and as they prospered and proliferated to satisfy a mushrooming domestic market and rising export commitments, there was a strong increase in the output of moderately priced yet often superb goods of all descriptions. It was a rare household in Europe and the United States, and ultimately in the world at large, that did not know one or another of this new breed of luxury and semiluxury items produced by Japanese industrial technology, and which went far to improve the quality of life for the average consumer who, perhaps for the first time, was able to afford such things. Meanwhile, the older industries such as textiles were easily surpassing prewar levels of production, but often in new directions. Textiles, for example, were becoming dominated by modern synthetics, while cotton production was suffering increasing competition from such growing industrial regions as India and Pakistan, Hong Kong, Singapore, Taiwan, and mainland China.

Revenues from increased sales of light industrial and such heavy industrial items as ships were used for the most part to create and improve heavy industry. This was accomplished in part through the banking system, although by the early 1960s the stock market grew in importance, conditions for it being enhanced by the housing shortage, the not yet mature development of the automobile industry, and by a general dearth of large-scale private investment opportunities at a time when incomes were rising and prices were relatively stable. At the same time, a high rate of personal savings — either indirect savings as before the Second World War, or direct as after 1954 — has been characteristic of the Japanese since the Meiji period, and an important contributing factor in economic growth.[4] It should be remembered also that in the postwar period Japan has been spared the debilitating cost of supporting a major military machine.

For this first period of postwar industrial construction it would be remiss not to stress the vital effect of the Korean War, both on the shift from social and political to economic growth, and on the decision to emphasise heavy rather than light industry. As in the First World War, the Japanese were called on only for material assistance, and while rendering valuable services for the United Nations' effort by providing innumerable industrial and other services, the experience was specially important in laying the groundwork and in providing capital for future heavy industrial development on an even greater scale than before. By about 1955 the directions of this development were well established. As has been suggested, the new industrial landscape, while reflecting past patterns, was frequently much larger and was often localised in new areas in response to unprecedented demands, both at home and abroad.

## The dual industrial structure

It may be appropriate at this point to assess briefly the roles of modern and traditional industry in Japanese economic development. Before the Second World War, differences between the two in terms of wages and the quality of the work force were probably less significant than at present. At that time traditional industry and Western-style or modern industry were both organised more strictly round familiar Japanese concepts of paternalism and patronage which emphasised loyalty and discouraged or virtually precluded horizontal mobility. Moreover the modern industries of the prewar era were not export-oriented, and there was little outward prosperity, at least that could be felt by the average employee. In any case, there is evidence to show that the two were probably more complementary than competitive.[5] Nevertheless, from early in the twentieth century the modern industrial sector, which was coming increasingly under the control of the *zaibatsu*, took the lead in wages and other benefits, and while there seems little direct testimony that this generally resulted, as it has since the mid-1950s, in a brain and manpower drain to the detriment of the smaller concerns and frequently in their outright bankruptcy, the differentials continued to grow until the outbreak of the Second World War.

Seen in this light, the term 'traditional' refers more perhaps to the administrative organisation of industries than to the kinds of goods produced, although there is a rough distinction on this basis. Typical traditional industrial products include cotton, silk and other textiles, toys, glass and chinaware, wood and paper products, whereas modern industries are more concerned with such items as heavy metal products and machinery, ships and other transportation equipment, electronic devices and petrochemicals; but there is a blurring of categories, especially when modern goods are manufactured by traditional small-scale industries, and to some extent vice versa. As already noted, the traditional sector has grown steadily weaker in the postwar period and today such industries are often directly controlled by their modern counterparts, being bound by subcontracting agreements and the like.

Beginning in the mid-1950s and especially in the past decades, the concentration of manpower and resources in the hands of large, modern, capital-intensive industries has been more and more along lines familiar in the West. Horizontal mobility, even among factory workers and especially among directorial personnel, has grown in significance, and the ability of small concerns to compete has been reduced. On the other hand, quite unlike the West, where the demise of a small company or business may be regarded as a 'natural' (if undesirable) consequence of economic circumstances, from which the victims ordinarily must extricate themselves as best they can, in Japan the practice is more for the larger and relatively successful to succour the failing, not only when the smaller companies are forced out of business, but in the whole process of modernisation and the creation of more efficient operations.[6]

The sixth and final segment of this breakdown of the march of modern Japanese industrial growth is the period from 1955 to the present, and is characterised as a time of rapid growth and maturation. Technological modernisation has been the main theme of this period and is motivated by a desire to narrow the technological gap in development, especially with the United States, or, in a way, to continue the practise of buying or borrowing foreign technology that can be traced back to the time of Perry, and even beyond. At the same time, since it has become progressively more difficult to purchase this from abroad, both because of wariness on the part of the sellers as a result of the vigour of competition from Japan, and because of acute price increases especially after the so-called 'oil shocks' of the early 1970s, the Japanese are striving with rising intensity and considerable success to create new technology of their own.

The principal reason for quick growth after 1955 was thus the gathering momentum of technological change, achieved conjointly by Japanese and foreign effort, or, more accurately, by rising expertise within the country in combination with a growing ability of the Japanese to purchase foreign knowledge. In addition, opportunities for new departures in industry were presented shortly after the end of the war by vastly altered economic conditions, particularly through such immediate postwar policies as demilitarisation and *zaibatsu*-dissolution. However fanciful these may have been, they nevertheless had the general effect of reducing economic concentration in favour of a system of intensive and relatively free competition. As a result some of the barriers to technological growth were removed. In reaction, rather quickly there arose many new companies dealing in the products of the new global technology, and certain of these fast grew to the point of becoming tantamount to postwar *zaibatsu*. Often these were the creations of farsighted individuals with salient technical skills, strong imagination and business acumen, while others were the continuations on a greatly enlarged scale, of previous enterprises, now infused with a new spirit. Examples of the former include the Sony Corporation and Honda Motors, while the latter might be typified by the Matsushita Electrical Corporation (*National* in Japan and *Panasonic* abroad), and by its two competitors in the electric field, Hitachi and Toshiba, while in other fields there are Toyota and Nissan Motors, and countless other makers of both light and heavy industrial products.

Following the Korean War, profits from which went far to support such new industrial development, and thanks to many factors, not the least of which was the willingness of western manufacturers to sell their secrets, the Japanese were able to accelerate their acquisition of new technology and to reduce the handicaps to their progress presented by prewar emphasis on empire-building and the maintenance of important military power. From that time forward large payments were made each year for new information, equipment, and expertise, and more and more joint agreements were made with foreign concerns. For a while foreign capitalised businesses and items utilising foreign technology earned a higher profit than entirely locally financed and development products.[7] In 1961, for example, it was calculated that products made with Japanese-developed techniques accounted for 53 per cent of aggregate sales, while foreign and jointly developed commodities made up only 34 per cent; but that profits from the sales of the latter were double those of the former, signifying among other things that foreign-tinged items were probably of higher added value.[8]

Furthermore, growth since 1955 underwent important changes in direction, not only from light industry to heavy, but also to a new private consumer-oriented heavy industry. This latter was based in strong measure on electric power production from such raw materials as imported, chiefly Middle Eastern, petroleum and even on that from local nuclear electric plants rather than, as formerly, on low-grade Japanese coal or on hydro-electric energy. For example, an entirely new heavy industrial empire built around petrochemical production has spread into many parts of the country, its installations becoming conspicuous landscape features in tidewater locations and its operations forming nuclei for large-scale, heavy industrial complexes (*kombinato*) for the production, in addition to items made from petroleum, of such things as power, iron and steel, and associated products and services. While the consequences of this growth in terms of environmental blight drew national and even international attention and finally called for extraordinary legislation which has since curbed some of the more obvious harm to the environment, the immediate effects on expansion were spectacular. The late Prime Minister Ikeda's (1960–64) famous 'income doubling plan' was realised in seven rather than ten years, and the nation was soon forging ahead towards even higher goals in the attempt to reach Western, notably American, levels of income and living for the average citizen.

## Major modern industries and landscape change

Until the fateful period of petroleum crises beginning in 1973, all industry registered a pattern of almost continuous growth, but, as suggested, the traditional industries grew more slowly than the modern sector. Light industrial output tended to decline in relative importance, and there was an unrelenting movement toward capital-intensive as opposed to labour-intensive enterprise. Annual investments in plants and equipment rose from about $2 billion in 1955 to multiple billions at the end of the 1960s.

The giant of the postwar era, and especially after 1955, was the iron and steel industry. Nearly all the prewar capacity was either destroyed or dramatically outmoded, so that it was necessary to construct new facilities almost from scratch. Most of these were located at tidewater so as to gain the advantage of immediate access to a supply of cheaply transported raw materials, the great bulk of which are imported from abroad, or even if indigenous would most likely be delivered by water carrier. However, although more than 90 per cent of the coal and most of the other ingredients of iron and steel are normally imported, Japan profits to some extent by being able to shop freely in the world market and thus to experience a measure of control over operating costs.

Integrated production from modern plant agglomerations is the rule, with ore being unloaded by belt conveyors and conducted directly to blast furnaces, followed by the transference of molten metal to converters and finally by finished steel being moved quickly to dockside bottoms where it is exported with minimal handling and delay. Generally, the new Japanese steel complexes contain some of the best and most efficient equipment in the world. For example, there is more production of basic steel by low-cost oxygen converters than is true of most industrialised nations. Likewise, the consumption of expensive imported raw materials has been contained by the introduction of modern continuous casting techniques and gas-recovery systems. The average ratio of coke consumed per million tons of steel produced has been characteristically low in relatively recent years compared with other steel-producing nations; as a result of these and other cost advantages, Japan has risen to world prominence in the field, both as a producer and exporter. In 1978 it was third in the world after the Soviet Union and the United States, but as an exporter it was first, with around 30 per cent of the entire world market, despite the serious inflation of raw material prices since 1973 which resulted in ensuing years in what the Japanese term a 'structural recession' in the industry.[9] Notwithstanding the ill effects of petroleum shortages and rising prices, after about 1975 Japanese industry in general resumed an upward trend in production, albeit at a more modest rate than in the feverish days before 1973.

Closely allied with iron and steel are industries for the production of machinery and these showed even more startling increases than the foregoing. Furthermore, after the petroleum crises of the early 1970s these tended to recover more strongly and rapidly, although with some exceptions. The shipbuilding industry, for example, which between about 1955 and the late 1960s built nearly as many ships in most years as had all of Japan's rivals (Great Britain, Germany, Sweden, France and Italy) combined, suffered particularly, in part a victim of its own success. Japanese ascendency in this field had resulted from a number of conditions, including technological innovation, especially in the area of prefabrication and rapid assembly, and in the application of unique designs and automated ship-handling systems. The elite socio-economic status of shipyard workers among Japanese labour was also an advantage, as was the attitude of labour unionism, which was far less rigid than that of the West in terms of craft demarcation rules. Hence, Japanese workers were more free to shift from one task to another in a way that might have been impossible in the West, and when shipyard workers became redundant, as happened after 1973, it was possible with relative ease to arrange for their retraining and re-employment in such industries as automobile manufacturing or in other industrial realms which have been especially quick to recover. These last include such precision products as watches and clocks, photographic equipment and supplies, and a large assortment of items in the electrical field, all of which have shown significant increases since 1975 in output and sales.

Advances in all categories of machinery have been especially great because output in earlier postwar years was generally negligible, while for certain products (e.g., television receivers, tape recording equipment, electronic computers, copy machines, and even most 35-mm camera equipment) it was non-existent. In 1978, based on an average of 100 for 1975, the index of growth for all machinery was 131.3. It was 16.5 in 1960.[10] Much of this reflects the remarkable growth of the automobile industry.

The chemical industry which, thanks to notable growth in the 1960s of petrochemicals, plastics and synthetic rubber, had developed into one of the leading industries of the nation, was hard hit by the oil crises, yet only some (photosynthetic materials and medical supplies) were thrown into serious recession. The petrochemical industry, however, was especially important in the early 1960s as a focal point in a vast Government scheme to create a series of embryonic industrial cities in various parts of the country, chiefly outside the established urban core — a topic that will be discussed further in the concluding chapter in a summary of regional and other planning through the years.

In contrast to the heavy industries and chemicals, the advance of light industries has been less noteworthy. Textiles, for example, have shown declines in certain areas for years (e.g., some categories of cotton, rayon, wool, and silk), whereas a few of these were giants of industry in prewar and early postwar Japan. Synthetic fibres and cloth, as might be anticipated, have, on the other hand, tended to fill the gap; none the less, the index of growth of the textile industry has been modest. It was 107.7 in 1978.[11]

Prominent modern industries also include food and allied products, where progress has been seen particularly in canning and preservation, even in freezing and such other modern applications as freeze-drying. The manufacturing of such Western alcoholic drinks as beer, wine (which has been accompanied by the rapid spread of viticulture) and whisky, for example, has proliferated along with increases in the production of such non-traditional foods as butter, cheese, mayonnaise, yoghurt and other foreign dietary items. Japanese eating habits have changed and more proteins and fats are recorded in the average per capita daily diet, though caloric levels (which have always been low by international comparison and were a subject of embarrassment) have not increased (they have been about 2 150 per person per day since 1965), but the matter is now probably looked upon with pride, considering the health-consciousness that has gripped the modern world. Indeed, by comparison, private economic, health and other standards in Japan, especially average life-expectancy levels, have become exemplary.

Other industries to enjoy the return of prosperity since 1973 are paper and pulp, ceramics, and especially the construction industry, with which the former is related. Let it be remarked parenthetically that ceramics and cement are also important articles of export. In cement production Japan is particularly favoured, standing second in the world in 1978 after the USSR.[12] As has been noted, construction in the strict sense of the term, though flourishing as never before, is seldom able to keep pace with the demands for housing. Some of this lag is attributable to sharp increases in land values, which limits the availability of construction sites. Chronically acute inadequacies in housing can be imputed to a variety of other causes as well. These range from traditional attitudes (e.g., tolerance of excessive noise, overcrowding, and similar inconveniences) to such modern trends as inflation and a conspicuous lowering of earlier standards of beauty and craftsmanship.

Another important industry which has experienced marked advances in this last period of postwar development is electric power generation. Japan has produced electricity on a wide scale for industry and private use since early in the twentieth century; even before the Second World War the use of electric energy by ordinary citizens was common and fairly universal. However, most of the power was created by water-driven turbines at isolated dam sites, the power being transmitted long distances by high-tension line, or by local coal-powered steam plants. The total amount of electricity produced was never sufficient to allow the nearly unlimited private consumption common in the United States and in many of the countries of northwest Europe. Per capita consumption of electricity in Japan has therefore always been less than that of the West. This situation persists even today, despite an enormous increase in total power brought about by (a) the gradual shift in emphasis from hydroelectric power to electric energy based on petroleum and coal; and (b) by the development of nuclear facilities, all producing cheaper, more available power.[13] The use of nuclear facilities has engendered problems similar to those in the United States and other countries but plans are going forward for the enlargement of this means despite uneasiness over mishaps, at home and abroad. Meanwhile, programmes for the furtherance of many often novel schemes to create electricity are being pursued vigorously. These include effort toward the perfection of coal gasification and liquefaction, solar energy, and a host of still unusual techniques, but nuclear energy is strong on the list of future alternatives. Fast breeder reactors such as are now in use (there were nineteen in 1979, producing some 4 per cent of energy needs, with nine more under construction and seven planned) are scheduled to provide the bulk of the nuclear power produced, at least until the 1990s, when it is hoped that the safer fusion reactors will have been rendered practicable.[14] It must be noted, on the other hand, that the number of plants in service, as well as plans for the expansion of nuclear facilities, tend to change or halt as a result of such circumstances as 'accidents' in existing plants (as in 1981 at Tsuruga Bay), or because of changes in international agreements over the reprocessing of waste materials, or even because of sudden oil strikes within Japanese territorial waters.

Most of the power in the near future, regardless of plans for the spread of other modes, will probably be produced by conventional means, but the share of electricity from nuclear and other forms of energy, is expected to increase. Forecasts for the mid-1980s, as of 1979, show the following as the major sources of electricity: imported petroleum, 66-72 per cent; imported coal, 11-12 per cent; nuclear energy, 6-7 per cent; LNG (liquefied natural gas), 5-6 per cent, and water, 3-4 per cent.[15]

## A new industrial landscape

Since about 1955, industrial expansion and its impact on the landscape seem to have been taking two different directions. It may be that these are destined to merge, at least in the present urban-industrial core region. The first such distinction has been an enlargement of previous manufacturing centres as a result of increasing and often excessive concentrations of industry within the larger cities. In such centres the machinery industry is strong, especially in the production of electrical equipment, construction devices, subsidiary equipment for industry itself, automotive products and other transportation items. New industrial 'estates' or 'parks' are common features of this development, notably in the metropolitan areas of Tokyo, Osaka, and Nagoya. These normally contain employee residences or apartment complexes, schools, playgrounds, shopping, medical and other facilities — all conveniently located with respect to factories, offices and to local transportation, particularly by private car.[16]

The other direction has been the development of large-scale industry in coastal locations, often on reclaimed lands and as a specific result of the national policy of accelerated economic development in the 1960s. The locale of these, as stated, is heaviest along the Pacific coast, particularly around Tokyo, Ise, and Osaka Bays and around the Setonaikai, where transportation is most convenient and large domestic and overseas markets are closest at hand. Petrochemical *kombinato* are especially conspicuous, along with complexes of other industries including synthetic fibres, gas chemicals, and iron and steel. Miyakawa recognises also what he terms 'Interlying' areas, in addition to the Setonaikai region, in this category, and he mentions Tohoku and Hokuriku of northeastern Honshu. There are also two 'outlying' industrial areas, both enlargements of prewar installations or outgrowths of industrial expansion in the 1960s. These are the industrial districts of southern Hokkaido and northern Kyushu. General reasons for the proliferation of industry in all these instances are said to be: (1) prewar evolution, (2) wartime industrial policies, (3) patterns of public investment after the Second World War, (4) postwar economic competition, and (5) the social organisation which supports what Miyakawa calls the 'oligopolistic' structure of industry in postwar Japan.[17]

With these developments have come changes in the character of many long-established industrial cities, even in remote locations. The city of Okaya, an example already given, has changed in emphasis from a centre for silk textile production to one of optical and precision instrument manufacturing. Ashikaga, another old centre of silk weaving, located in the northwestern outer reaches of the Kanto Plain, has retained its prominence in textiles but has since turned to the production of knitted tricot materials in modern industrial 'parks'. Yokkaichi, situated on the northwest shore of Ise Bay in Mie Prefecture near Nagoya, is a third example with a background in textiles, which now has become one of the nation's earliest and most famous petrochemical centres, its waste products earning it in the early 1970s an infamous reputation as a major contributor to the pollution of offshore waters and of the atmosphere in the Chukyo region. So severe was this that subsequently, thanks to stringent legislation which came about in large part because of conditions in this area, there has been massive improvement, particularly in air pollution.

7.3 An extensive area of land newly reclaimed for industrial use along the foreshore of the Seto Inland Sea (Setonaikai) on the eastern outskirts of Hiroshima

Petrochemical, steel and other heavy industrial complexes, as noted, have spread throughout the whole core region, being of special importance in the Tokyo Bay area around the huge industrial cities of Kawasaki and Yokohama, and in the vicinity of the city of Chiba. Similar development is seen in the Osaka region round the ancient port city of Sakai as well as in and around the neighbouring cities of Wakayama to the south, and of Amagasaki, Nishinomiya, Kobe (where an entirely new 'port city' has been constructed on reclaimed land) and Himeji, on the Harima Plain, to the west. The Setonaikai region is punctuated by such centres, as in the South Okayama (Okayama Kennan) district, a part of the city of Kurashiki, with similar growth in many places in southwestern Honshu and in northern Shikoku, even in eastern and southwestern Kyushu, between Oita and Nobeoka, and around the eastern shores of the Ariake Sea.

In the old Tokaido region, between Tokyo and Kobe, where both kinds of development have occurred most intensely, the pace of change has been so furious that the entire region is seen to have become one gigantic, virtually unbroken metropolitan area, known, as mentioned in Chapter 1, as the Tokaido Megalopolis.

Areal development of industry in the 1960s was furthered also by support from the central Government in the form of treasury loans and investments which were much increased as time went on, the scope of such financial support being enlarged to include nearly the entire nation. For example, in certain places where major construction was contemplated, small landholders were sometimes persuaded to sell their rights by means of combined pressures from the central and local Governments in concert with those of private industry.[18] Core elements of the new industrial project areas, whether they were Government-sponsored or under control of private corporations, were often carefully planned and included ancillary facilities with a view to creating conditions of optimum efficiency and at the same time certain comforts for the residents. But outside the main planned areas there often arose a mixture of decaying agricultural land and haphazard, unplanned construction.

In local centres where industrial reorganisation and modernisation also occurred in somewhat the same fashion but perhaps more slowly and on a smaller scale, major factories were frequently surrounded by related subcontracting plants and other facilities, the whole being well integrated and often logically arranged in terms of a smooth flow of materials but perhaps lacking in aesthetic appeal for those who live there. The automobile industry best exemplifies such growth, especially around the city of Toyota (formerly Koromo), in Aichi Ken, Japan's main automotive city.

Finally, it should be reiterated that developments of the kind noted in the latter part of this chapter greatly intensified the problems of 'public nuisance', the literal translation of the now popular term *kogai*, a euphemism for environmental pollution. Japan came closer perhaps to having its physical environment damaged or significantly altered than is true of other industrial cultures because of a dearth of space, the speed and weight of modernisation, and until rather recently, a lack of public awareness of the problems — this last being a condition common to all 'developed' societies. Manifestations of this situation have been apparent for many years and in the postwar era there have been a growing number of studies and attempts to use technological and legal means to alleviate the difficulties. Two problems to come under particular scrutiny within the past two decades are atmospheric pollution and ground subsidence, both of which are most pressing in the larger urban areas. The environs of Tokyo, Nagoya, Osaka, and Kita Kyushu, for example, have been especially plagued by contamination of the air, so that only by the most extreme measures, including unusually tough legislation to curb the discharge of effluents from factories and cars, has there been improvement in recent years. The use of private vehicles, which has expanded relentlessly, has called for special restrictions in crowded areas of Tokyo and other cities on certain days, and laws have been enacted to promote the use of propane as fuel for taxis and other conveyances. Subsidence, with an even longer history, is accredited to several presumed causes, chief among which are overuse of ground water as a result of heavy industrialisation and excessive population densities (especially when associated with reclaimed lands as in Keihin, Kansai, and other places) and the extraction of natural gas, as in Niigata. To these annoyances may be added longstanding problems of general crowding, noise, vibration and befouling by chemical agents. In the last case, there was much public clamour for

*Summary*

7.4 Sunday on a main street in the northern part of central Tokyo when automobiles are excluded by law

restrictions against chemical pollution of offshore waters, a situation that was spectacularly punctuated by several fatalities resulting from direct contact with poisonous wastes, either in the atmosphere or through the consumption of diseased fish. Newspapers continue to give almost daily accounts of dangerous substances in milk, vegetables, and other foodstuffs.

## Summary

Japan's new industrial landscape is more extensive than before the Second World War, although its central core is essentially that of the past. At the same time, there has been sizeable growth of new industry, especially in the coastal parts of Tokaido, Chugoku, Setouchi and northern Kyushu, with important and unprecedented industrial agglomerations even in Tohoku and Hokkaido. Only the coasts of the Sea of Japan, westward of the Noto peninsula, those of southern Shikoku and of northern and eastern Hokkaido are without significant industry. Even in some of these areas, however, development is contemplated.

Japan's reputation as an industrial nation has been reversed. Whereas the Japanese were once held to be devoid of originality or were thought to be concerned only with profit or with military production (which was of proved effect in the Second World War), they are now renowned for quality production not only of light and precision goods for the international carriage trade, but also for a full range of heavy industrial

141

products from basic steel and ships to reliable and well-designed motor vehicles and other sophisticated capital items.

Industry has become so pervasive in recent years that opportunities for non-farm employment, in all but the most remote parts of Tohoku, Hokkaido and southern Japan, are available on a daily commuting basis. Able-bodied farmers may still live on the family farm, but their incomes are predominantly from other pursuits.

The first postwar period, one of reconstruction, began in the late 1940s and saw the return of certain prewar industrial specialities such as textiles, and the initiation of many new industries, for which the Japanese reputation is now often supreme in the world. Growing cold war disillusionment on the part of the United States and the West, especially after the Korean War, helped to turn the reconstruction of Japanese industry from an emphasis on light goods and precision products, to heavy industry in the fullest sense of the term. And in the decade or more during which the new, ultramodern facilities for producing petrochemicals, ships, steel and other goods were being constructed, light industries expanded, grew increasingly inventive, and usually underwent consolidation in the hands of fewer and fewer large companies catering to an ever widening market abroad and especially within Japan. By the early 1960s, thanks to rigid inspection standards and effective after-service policies, the old anti-Japanese sentiment against shoddy quality tended to fade and die out, even in the United States.

Modernisation has much increased the importance of the capital-intensive over the traditional, labour-intensive industries, encouraging more horizontal mobility among workers. This has resulted in a brain and manpower drain in favour of the modern industrial sector. On the other hand, the dichotomy in question remains a hallmark of Japanese industry, especially as there has been a tendency for the larger and more affluent industries to assist those in difficulty.

The final period, that of rapid growth and maturation, has therefore been dominated by heavy industrial development, partly as a result of association, especially with foreign concerns and business interests. Industrial innovation, considered the cornerstone of the spectacular growth in the 1960s and early 1970s, has come from both indigenous sources and abroad: the pooling of domestic and foreign talents and other assets has been particularly advantageous to Japan in terms of its overall wealth and economic position in the world. The price which Japan along with other nations is now paying is environmental pollution and a general deterioration of the 'quality' of life. Individual living conditions have become even more cramped and difficult than in the past, to the point, in many areas, of endangering public health, but there has been some amelioration of conditions since about 1975 as a result of special legislation stemming from public outcry. The pace of life has become ever more hectic, despite salient increases in personal monetary income.

The major postwar industries are petrochemicals, iron and steel, the

# Summary

machinery industries, ceramics, paper and pulp, and construction, but all industries suffered setbacks following the period of 'oil shocks' in 1973 and it has been the ones mentioned above which have tended to regain economic health most quickly and completely. Shipbuilding, which in the world was almost an exclusive Japanese endeavour between about 1955 and 1973, because of various circumstances after 1973 has suffered particularly, together with light industry in general, while those which depend upon popular demand, such as electric power and food, have managed to hold their own. Petrochemicals exemplify the industries that have evolved entirely since about 1955, and in this particular case the effects on the growth of new thermal and nuclear power facilities and on agglomerations of various heavy industries, even in the outlands, are noteworthy.

The modern industrial landscape is thus larger and more pervasive than before. On the one hand, industrial cities are rather generally spreading their influence into their hinterlands; on the other, new sites are being created along the coasts, especially of southern Honshu, the Setonaikai, and Kyushu. Machinery manufacturing is common in the former, and petrochemicals and related products in the latter; in some areas (e.g., parts of the Tokaido corridor) the two may be indistinguishable. Plans for new industrial centres are often well laid and executed, thanks to cooperation between Government and private industry, but outside the planned areas a somewhat chaotic situation may exist. An example of a carefully planned and effectively operating milieu of new industry is seen in the automotive region around Toyota but conditions for those who reside there may be less than satisfactory.

Public nuisance, a problem which grew at an alarming rate until the early 1970s and is still far from conquered, embraces pollution of water and air, and includes such age-old conditions as excessive noise, vibration, and ground subsidence, all of which are most extreme in the major urban-industrial core.

# 8
# The Japanese landscape

## Natural setting and human adaptation

It may be seen in the foregoing chapters that from the beginning of history until quite recent times the Japanese landscape has been dominated by rural activity. It has been and to a great extent remains a landscape of remarkable beauty, the 'natural' qualities of which have in many ways been fashioned by man in his efforts to raise ever larger amounts of rice and other products. If anything, the Japanese have improved upon nature by reducing their landscape to a kind of order, by investing it with an aura of 'studied naturalness', a characteristic that has long been associated with their artistic expression, as exemplified in the arts of garden architecture and the raising of dwarfed trees (*bonsai*).

From the earliest times the Japanese have felt a closeness of association with nature, a posture deeply rooted in their original culture and later reinforced by Chinese attitudes taken in with such philosophies as Taoism, Confucianism, and Buddhism, especially as conveyed in Zen. These included a respect for natural phenomena, manifested as a spirit of cooperation, of preferring to retain a balance between feelings of gratitude on the one hand for nature's benevolence, and of awe on the other, implying the acceptance of nature's probable dominance — as was constantly being suggested in periodic and frequent natural disasters. It has only been in very recent times that widespread non-agrarian modifications of the natural landscape have been made. Even so, the present may be merely another transition leading into a new era of *accommodation* which probably more than any other term signifies the means by which the Japanese have attained success in managing their environment.

As a setting for human development, the Japanese Archipelago has been in many respects a hostile if not often a violent land. While man has co-operated with nature in creating a rural landscape that appears both rich and tranquil, life for most Japanese has been hard throughout their history, for the achievement of cultural, political, and economic unity in this beautiful but stark environment has entailed a heavy price. To do more than merely survive in such a land has required all the strengths and ingenuity of a resourceful people. Indeed, down to the very present, generation after generation, in addition to suffering the usual hardships inherent in the agrarian life, has sacrificed pleasure, gain, and security for the advancement of society as a whole.

And now at long last the 'rewards' have begun to appear, broadly distributed throughout a population long inured to rigorous self-discipline and self-denial. Through creative human management the nation has raised itself to the status of an industrial superstate, and the new period of mainly capital-intensive, machine-oriented mass production into which Japan has now emerged, has introduced the public to higher levels of existence than it has ever known before, showering down on the common man a rich assortment of material possessions. Japan is now on the threshold of affluence, thanks to wise leadership, good fortune, and especially to the patience and endurance of the people. In the final analysis, the latter have constituted Japan's richest natural resource, and considering the woeful lack of other resources, their accomplishment has been heroic. The time is not far off perhaps when such hitherto alien concepts as individualism, leisure and privacy will become logical human goals. Whatever the consequences, Japan's achievement is an irrefutable instance of the triumph of the human will over environmental limitations.

One contemplating the obstacles the Japanese have overcome pauses to ask some open-ended questions as to what results when centuries of hardship culminate in affluence. Was the severe discipline to which the Japanese were constrained really the 'price' paid for what they have now achieved, or did that discipline constitute its own 'reward' in its own time? Is the advent of material prosperity the 'reward' of their experience, or a 'price' of another sort that spells the end of everything we admire in their make-up?

One of the consequences of an abundance of rich human resources has been the sustained transformation of the landscape from an agrarian setting supporting a growing non-agrarian milieu, to the very opposite. In the present day, therefore, traditional ways are fast being replaced, and though the evidence in the landscape may often be elusive, statistics show that most Japanese are urban, at least in terms of income, outlook, and to a surprising extent when compared with the history of the city in other cultures, even in their heritage.

## The agrarian landscape and agriculture

In the face of these aspects of modernisation the mainstay of the economy continued for years to be primary production, and until the 1960s non-agrarian landscape changes were largely superficial. Long before the end of the feudal age, agricultural practices had become highly sophisticated. Based on techniques acquired before known history and supplemented by influences from Korea and China, agricultural progress has been constant, reaching particular heights in early Tokugawa times, later under American influence, in the early Meiji period, and finally, since 1945. Throughout, progress has been particularly noteworthy in such areas as water control

*The Japanese landscape*

and soil management, but there have been advances as well in the whole area of agricultural resource conservation. Emphasis has constantly resided in the cultivation of irrigated rice, now near urban centres supplemented by such as truck crops and flowers (the old off-season reliance on dry grains having been largely abandoned as uneconomical) and year-round production where possible of a wide variety of vegetables and fruits. Protein has been derived mainly from aquatic sources, notably the sea, but also from such as soya beans, to some extent from eggs and chicken (this last accelerated recently by changing tastes) and also in the modern day from dairy products and meat, particularly pork. The narrow alluvial lowlands from which most of the food has come have steadily been improved and enlarged but near cities some has been sacrificed to expanding urbanisation.

On the other hand, the Japanese peasant was traditionally slighted in terms of economic and social rewards, despite official lip service to the importance of agriculture. The organisation of society, cemented into the culture by centuries of practice, was of a nature to keep the peasantry in subservience, while the rest of the population relied on its services. In the famed words of George Sansom, 'Japanese statesmen thought highly of agriculture, but not of agriculturalists.'[1]

Consequently, until the momentous land reforms of the Occupation period, the Japanese farmer rarely held title to his land, was chronically in debt, subsisted on a diet that was less appealing than that of city folk, and generally had a prearranged style of life from which there was little

8.1 Large new prefectural hospital near the border of Shizuoka and Kanagawa Ken and typical of such modern facilities in or near major cities. It depends mainly on private drivers or buses

opportunity to deviate. Yet, also a product of traditional values, the farmer usually endured his lot and passively accepted his role as primary producer, even rising above the desire to redress his grievances by devoting himself to hard work and the constant improvement of his surroundings. In the main his life was one of restraint, and although history records sporadic peasant riots of some magnitude, there have been no large-scale popular revolutions.

The postwar period has seen the disappearance of many of the foregoing conditions. Absentee landlordism, payment in kind for rents and taxes, oral contracts and other forms of subterfuge that worked against the security of the farmer, are all of the past. As a social class, he now suffers little discrimination, even from intellectuals; he has probably tended to remain at a relatively modest level in terms of income, either because other industrial sectors have been more prosperous or because his further professional development has been limited by such age-old and almost insoluble problems as a lack of cultivable land or the fragmentation of his holdings. The subdivision of agricultural lands for new housing or new industry, and many other manifestations of spreading urbanisation, have been lucrative for those whose properties lie in the path of such expansion, but such trends are also indicative of a general retreat of agriculture in the present day. Legislation to curtail rice production has also helped to cōntain major agricultural development in certain areas. The Hachiro Lagoon of Akita prefecture is an outstanding example of such legal constraint. This area of more than 22 000 hectares, about two-thirds of which, expanding from lake shore areas reclaimed before modern times, in the early 1960s was converted into paddy fields to become a new area for large-scale rice production by the most modern techniques. Workers were specially selected from other places on the basis of their knowledge of advanced technology, the whole programme taking many years and expending large amounts of capital — costs being born mainly by the central Government, with some assistance from the prefecture. The latest equipment was introduced, modern housing and other facilities were provided, and elaborate water-control measures adopted. But on the eve of full-scale operations the Government decided to legislate a nationwide reduction in rice acreage, and suddenly the future of the area was cast in doubt, except for various suggestions, among which was the proposal to turn most of this rather vast tract into another industrial complex, a course which would probably displace many of the residents, present and prospective.

However, it may be misleading to discuss the lot of the Japanese farmer *per se*, for, as repeatedly suggested, his primary income activity in recent years has shifted steadily to non-agrarian occupations. In regard to his overall life style, he may be attuned as much to the new style of living as to familiar rural patterns, while in his beliefs and attitudes he may often be closer to those of his forefathers than to those of more long-established city folk. In short, the present is a time of rapid transition, although indeed it may be that the term 'farmer' has never been appropriate for the

Japanese worker of the land: 'horticulturalist' may be nearer the mark. It would be equally incorrect to think of the modern Japanese rural resident as an urban person, although in many ways his life and values are decidedly a part of the contemporary urban tradition. The typical Japanese of the future will surely be non-agrarian, but whether he may be considered urban in any other than a purely Japanese context, is a moot question. On the other hand, we may be reminded at this point that, while the term 'civilisation' is etymologically an urban concept, the term 'culture' is all-embracing. It is obvious that in the writer's opinion, the agrarian Japanese has always had in his basic make-up an unusual amount of urban culture.

### The urban landscape

Japanese city culture was one of the consequences of increasing Chinese influence after the seventh century. While its spread was slow, taking nearly nine centuries to become a significant style of life, most Japanese grew familiar, from the literature or hearsay, with their imperial or aristocratic world which, if not entirely urban in our eyes, was certainly divorced from a primary stage of existence. In time, as cities became more important, a growing number of commoners came to adopt, through trade, commerce, manufacturing, and the services, the non-agrarian life, and there was, between the fourteenth century and the seventeenth a long period of growth in all these respects. Urban society continued to be dominated by the military and imperial aristocracy, often in concert with the clergy, who operated chiefly from headquarters at Nara and Kyoto on the one hand, and from various military strongholds on the other. By the early sixteenth century, however, life in the cities, although still controlled by the aristocracy, was becoming increasingly proletarian – a tendency that gathered momentum with the initiation of Tokugawa rule in 1600.

Several kinds of towns can be identified by function, but gradually the castle town (*jōka-machi*) became the focal point of non-agrarian activity. Mostly these were local command posts, but from about 1575 their immediate site characteristics, chosen not only for strategic advantage but also for ease of communications and especially as centres of productive hinterlands, were often favourable for the growth of population and even for subsequent modernisation. From the seventeenth century onwards, the city assumed a particularly vital role in the vigorous and almost wholly indigenous commercial life of the long Tokugawa era. Castle towns were, in fact, the fulcrum of the economic organisation of the state, and in terms of the sizes of cities and of the interdependence of their relationships, the complex hierarchy of settlements, carefully linked by well maintained, closely guarded routes of travel, which provided the framework for the dominance of Edo, the supreme military capital; for its size and scope and within a single unified culture, may be one of the earliest examples in human history of such a phenomenon.

*The urban landscape*

The Tokugawa network of cities was, of course, a product of a firmly autocratic, feudalistic governmental system. After the collapse of the latter in the 1860s, there was a general reorganisation of society and, unlike prior experience, the driving force was in many ways more nearly akin to Western concepts of *laissez-faire* capitalism. None the less, the period between 1868 and 1900 was a time of controlled growth, and the combined efforts of the forces of tradition on the one hand, and of a modern, non-Japanese element, on the other, produced rich and rapid rewards. Before the turn of the twentieth century, Japan had entered the arena of international power politics, and the landscape had begun to reflect changes away from a purely agrarian base.

Transportation had been modernised, with railways superimposed on the old highway pattern. New ports to handle international shipping had been constructed, generally between Kanto and northern Kyushu, in response to increasing United States and European trade, and at least two of the new outports had moved into the top echelon of cities. Hokkaido settlement, including urban and industrial development, was well under way. There were certain new city types, based on such unprecedented functions as modern strategic needs, heavy industry and transportation, though most places of importance after 1868 were outgrowths of Edo period centres. By 1900 also, the Tokaido and Seto Inland Sea regions, including northern Kyushu, had begun to show comparatively greater vigour in attracting migrants from other areas, and the cities of these

8.2 Neighbourhood department store, Tama New Town in the western suburbs of Tokyo but close to Kawasaki and Yokohama. The store is one of a chain run by the Tokyu company which also operates extensive commuter rail lines in the Kanto district. Note family in the right foreground

## The Japanese landscape

regions, even more so than in the past, were becoming the most prominent in the land.

After 1900 this tendency steadily grew and was augmented by industrial and commercial expansion, favoured especially by wars. And even though all major centres were destroyed in the Second World War, the Tokyo–Osaka urban axis and its extensions along the Pacific and Inland Sea littorals have since dominated the urban scene more than ever. At the present time the prefectures between Kanto and Kansai contain roughly 65 per cent of the total population, and the rate of growth consistently rises. As in the West, the population of this urban region has grown with particular rapidity in the areas between cities, with the cities tending to stagnate or to lose population as their natures change from the more traditional pattern of combined work and residence, to the Western style of being chiefly work places by day and combined amusement centres and ghost cities by night. As a result, urban sprawl has become a problem of intense importance in contemporary Japan, with the usual side-effects of industrial and human pollution. National affluence, which has encouraged the average citizen to amass even greater quantities of costly material possessions, and more recently, to invest in the long-term purchase of individual residences and family cars, has compounded the issue so that the Tokaido Megalopolis has become one of the most severely crowded and cluttered regions of its kind in the world.

### The industrial and commercial landscape

While industrial landscapes of a scale large enough to dominate an entire subregion are comparatively recent in Japan, in a broad sense the Japanese have long known such processes as centralised production, subcontracting and assembling, organised commercial exchange, apprenticeship systems within guild structures, industrial or manufacturing specialisation (often in conjunction with raw material exploitation), and trade of notable proportions, both domestic and foreign. Craftsmanship of matchless quality was common even before medieval times, especially in the making of such famous Japanese products as swords and armour; wood, leather, paper and lacquer goods; ceramics, textiles, and other wares, many of which have become increasingly valuable in an artistic sense. There was also commercial manufacturing of such foods or beverages as soy bean products, vinegar and *sake*. Proud and careful workmanship was likewise characteristic of the architectural and building trades, of general woodworking, or iron and other metal manufacturing, and of shipbuilding.

During the long Edo period, internal trade in rice and other foods as well as in manufactured goods helped to widen the scope of commercial life and of premodern industry. Foreign influences were few, but were seen in the making of firearms and in certain techniques of construction, especially of castles. Particularly noteworthy was the experience gained at this time in business organisation, which resulted in the formation of large,

## The industrial and commercial landscape

often family enterprises, and in the general elevation of commerce as a profession. The Edo period also saw the beginnings of the enlargement and standardisation of industrial production in specific regions, and at times under the strong control of guilds or local monopolies. Tokugawa period developments in industry and commerce were of special importance in so far as they provided the framework and established operational procedures which were vital to later modernisation.

Japan's 'industrial revolution' came about, however, in the latter part of the nineteenth century, and development was slow until the Meiji Restoration of 1868. Nonetheless, there were important though limited moves toward industrial modernisation well before this date, especially in the fields of iron and weapons production, textile manufacturing, and shipbuilding; specifically, these took place among certain enlightened *daimyō*, notably in the Mito-han of northeast Kanto, and in several fiefs in far southwestern Japan. Concurrently, and equally important in helping to set the initial course of modernisation, was the practice of employing temporary resident foreign managers and advisers, and of purchasing foreign equipment and technology. The latter two expedients have been especially useful for rapid industrial growth in the period since the Second World War.

Large-scale industrial modernisation, the main lines of which were laid down within forty or fifty years after 1868, was accomplished through a combination of circumstances. Some of these were fortuitous, but most were the result of astute leadership of the common people, whose responsiveness enabled the nation to surmount their lack of capital, of resources, and of experience and technical knowledge.

Early industrial growth proceeded from an emphasis on light industry, especially textiles, of which silk was a most important international trade item in providing revenue for further industrialisation, including increasingly heavy industry for the construction, mostly for domestic use, of capital equipment and for the making of chiefly defensive armaments. Later, wars were a significant propelling force in industrial growth. This is especially true of the First World War, which for the first time in modern history saw the Japanese becoming prominent in the industrial export field, a development which by 1919 had reversed Japan's economic status from debtor to creditor nation.

The 1920s were marked by the rise of Japan in the textile industry and by the first signs of technological maturation, notably in the creation of original mechanical devices that were adopted by the rest of the world. At the same time, economic depression, especially in rural Japan, tended to nullify the optimism that had been engendered by a lively international exchange of goods and information; in the 1930s Japan turned to military production in order to support its growing ambitions in eastern Asia, and to the exporting of sundry items, more often than not of shoddy quality. Much of the export trade was with the United States which, conversely, was supplying Japan with the bulk of its strategic imports, and the result-

## The Japanese landscape

ing trade imbalances helped to precipitate war in the Pacific, and to close an era of great industrial and commercial expansion.

Japan's dazzling rise to greatness since the Second World War is especially attributable to industrial growth and to changes in the directions of industrial development, and there have been significant alterations of the landscape as a result. Secondary and tertiary industries have been particularly affected by post-1945 developments, contributing the lion's share of the sustained and unusual growth of the gross national product.

The extraordinary expansion of Japanese industry was also the result of such immediate postwar Occupation policies as efforts to dissolve the *zaibatsu*, the encouragement of labour unionism, the reorganisation of the political structure, including the enlargement of the voting franchise. Sympathetic economic attitudes, notably by the United States, were of great importance especially at the beginning. But equally noteworthy are many factors long common to Japanese industrial modernisation. These include enlightened leadership, the support by the central Government of private capital formation and growth, a general willingness by management and labour to work together in the national interest, and such traditional assets as paternalistic relationships within a modern industrial structure. Also, as in the early Meiji period, events in the world at large have been manipulated in such a way as to contribute heavily and favourably to Japan's success.

While the foundations of the modern industrial landscape are basically

8.3  Car ferry in the port of Sendai, August 1980. Services by such ships are commonplace for transporting passengers and their cars to distant locations. This line runs between Sendai and Hokkaido

those of the prewar era, there are important differences. The old industrialised suburbs and enclaves have all expanded, forming urban-industrial regions. In the central areas these are virtually unbroken by non-urban activity and its effects. In such places even agrarian pursuits are closely geared to urban demands for such as truck crops, fruits and flowers. Indeed, often nearer the larger cities, agricultural life is now ephemeral, the landowners merely biding their time, hoping for an increase in land values and a removal of restrictions, before selling out to developers and speculators. As in other parts of the world, the latter are frequently able to exploit loopholes in existing legislation in order to convert farmland and even attractive hill land in ever-extending suburbs into tracts for new industry and especially for housing. Such development in certain parts of the urban core (eg, Kamakura and the Kobe area), has even come to encroach on protected sites in the public domain and often of primary historical importance. Hence, the central and local Governments, pressed by irate citizens, are moving to exert greater control over the processes of expansion.

As a final note in this summary of Japanese industry and its position in modern cultural development, it seems appropriate to re-emphasise the change — a further index to the attainment of technological maturity — in the image of Japan as an industrial power. The Japanese are now regarded by their rivals in the West as competitors in all fields of modern industrial endeavour; in certain ones (eg, shipbuilding, photographic and other optical equipment, and in certain categories of electronics and perhaps in the automotive field) they are almost in a class by themselves.

## Urban, metropolitan and regional planning

Contemporary urban planning in Japan dates from the initial city planning law of 1919. Nara in its original form and Kyoto had both been built on plans imported from China in the seventh century. Influences akin to those seen in these early achievements can possibly be recognised in cities of later date, but the cities of the Edo period and of early Meiji times were largely unplanned, except perhaps for castle grounds and for certain gardens which have since become public parks and recreation areas. There may also be some hint of planning in the rectangular street patterns surrounding the castle compounds in many local cities, reflecting ancient defensive arrangements of moats or streets.

An opportunity for implementation of a modern, relatively large-scale plan for Tokyo and other Kanto cities was presented in 1923 by the disastrous earthquake in this area. Important changes were made when cities, especially Tokyo, were rebuilt. Streets were generally widened; zoning ordinances, particularly those aimed at fire protection, were strengthened; and building codes were introduced. But in the ensuing years, pressures exerted by immigration and burgeoning traffic tended to offset attempts to pursue a logical course of development; in time, the picture, all too familiar elsewhere, was one of random growth except in very limited directions.

In the 1920s, nevertheless, most prefectures and cities organised planning agencies which developed legal and practical machinery for the design and execution of specific programmes. These were either carried out gradually (without much regard to popular reaction) or were delayed until after the war. As elsewhere, this prewar planning consisted mostly of replanning, or the revising of archaic forms. Until recently, massive amounts of capital were rarely at hand to fashion important changes in cities. Such funds were made available only in cases of dire emergency or, in lesser amounts, for such public and semi-public works as parks, playgrounds, schools, wholesale markets, docks and other harbour installations, and for public transportation. City water was generally provided from suburban sources of supply, and since some of these facilities were modern and efficient, the quality of piped water was usually high by world, and particularly by Asian, standards. Gas and electrical facilities and telephone services were also common in cities, but comprehensive sewage systems remain inadequate to this day, in part because human waste was traditionally collected for use as fertiliser; although effort is now being made to enlarge underground networks for waste disposal, the open sewer is still a familiar urban and suburban feature.

The devastation wrought by bombing in 1945 presented a new opportunity for widespread, planned reconstruction of most Japanese cities. Unfortunately, funds for a comprehensive programme were lacking because of other, more fundamental and pressing needs. In addition, many plans were frustrated by squatters and local property-owners, against whom laws for the designation of space in the public interest were inadequate. Only in certain places (eg, Sendai, Nagoya, Matsuyama) were community-wide, modern plans initiated with vigour and imagination immediately after the war, and ordinarily these were products of unusual political control on the part of local leaders, coupled with uncommon foresight, public cooperation and other fortuitous circumstances. During this period of recovery plans were sometimes allowed to succeed in local cities because such outside pressures as were disrupting plans in Tokyo and Osaka, were weak and hence amenable to strong civic action.

These remarks apply also to metropolitan and regional planning. These may be said to be of even more recent date unless we choose to take account of Tokugawa efforts to regulate the growth of Edo, and other measures, then and in early modern times, all of which required planning and concerted direction by a central authority. It has only been since the Second World War, however, that large-scale modern planning on the metropolitan and regional levels has been pursued, academically and in the legislative sense. Comprehensive metropolitan plans have involved, first, the Tokyo area, for which the National Regional Capital Development Law, an outgrowth of previous plans and projects, was established in 1956, transferring jurisdiction from the metropolitan to the regional levels.[2] This

*Urban, metropolitan and regional planning*

was followed by the Hanshin Metropolitan Area Plan of 1960-62 for the Osaka-Kobe area and later by the Chubu Region Development Law for Nagoya and vicinity. Through the years all these have been revised and updated, sometimes to include other cities, such as in 1965 with the enactment of the Kinki Region Development Law, which drew Kyoto into the aforementioned Hanshin Metropolitan Area Plan.

The postwar period has thus seen a rapidly growing need for planning, not only for cities and urban areas but on a national scale. The first demonstration of this was the Comprehensive National Land Development Act of 1950, although during the 1930s there had been a planning board, whose organisation and purposes reflected various influences, notably

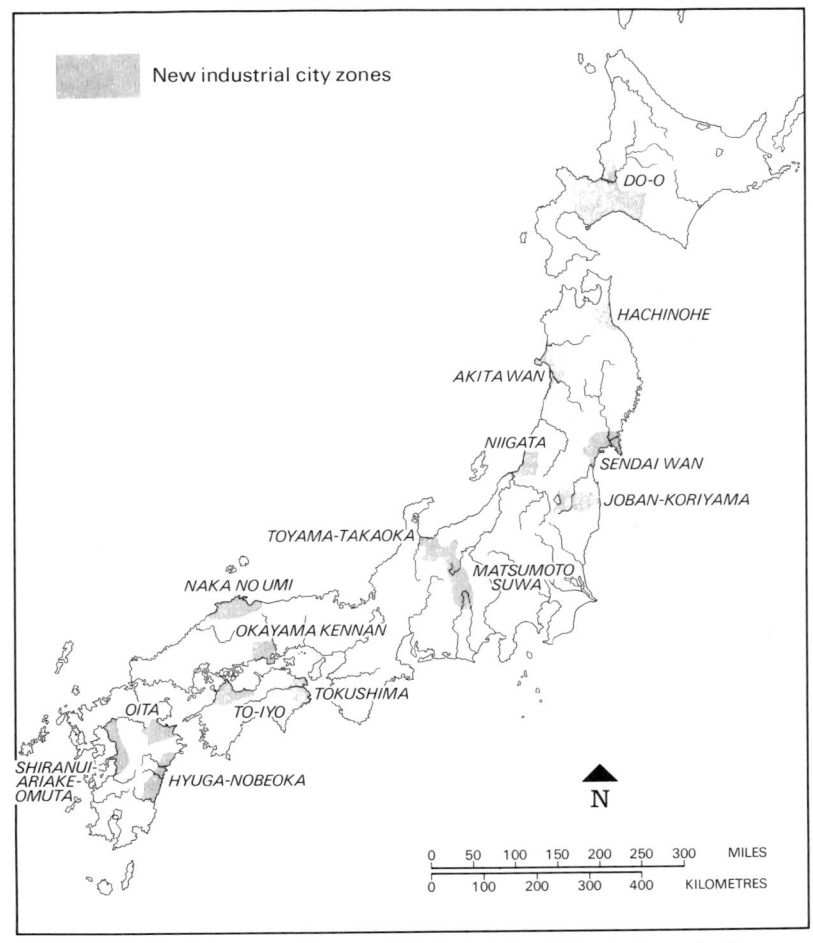

Map 13 Industrial city zones. (*Source: Shin Sangyo Toshi No Genkyo*, 1967)

## The Japanese landscape

those then in vogue in Germany. Action after 1950 was hampered by a dearth of capital, and it was not until 1962, on the heels of Ikeda's famous Income Doubling Plan, that the first Comprehensive National Development Programme was established as the cornerstone of future national development planning — a concept which had to be abandoned after 1973 because of oil shortages and subsequent inflation. One of its initial aims was to establish 'growth poles' of industry and settlement, particularly in the outlands. The Law for the Establishment of New Industrial Cities (1962) was a focal point of the plan but its implementation tended to lose momentum except where the 'growth poles' were in

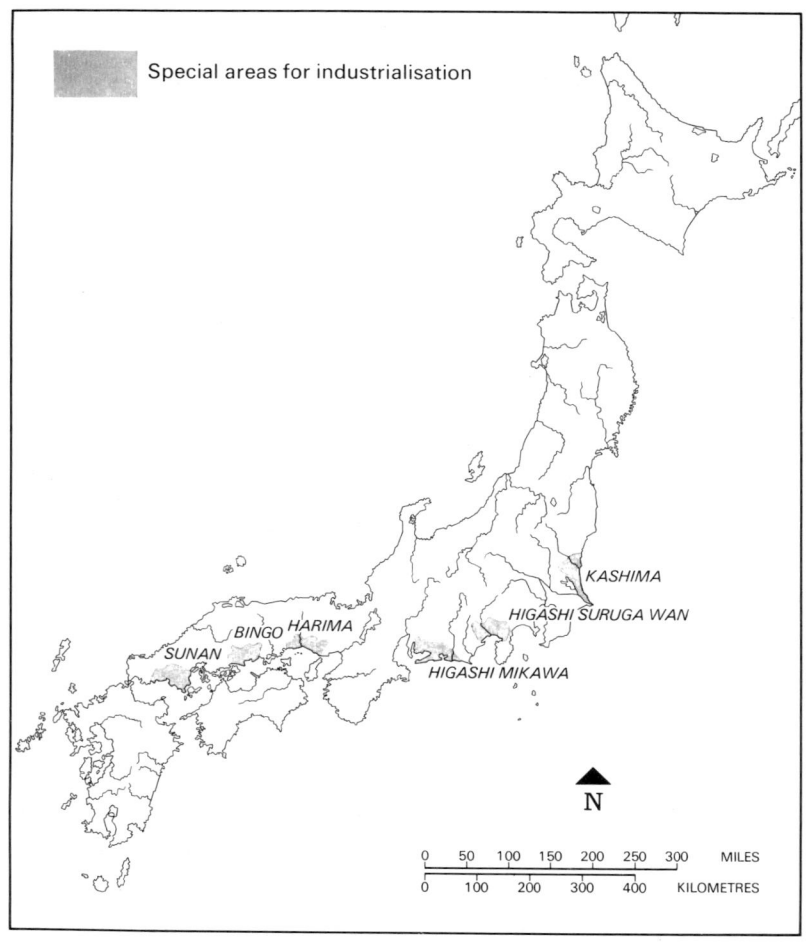

Map 14 Six new special areas for industrialisation. (*Source: Shin Sangyo Toshi No Genkyo*, 1967)

or near the urban core, thus weakening the programme in terms of its basic intention to spread industry and ultimately to slow population influx to the great corridor between Tokyo and northern Kyushu. A basic activity in each of the new locations was the swelling petrochemical industry and the plan attempted to set up, mainly in underdeveloped areas, a group of industrial cities and later several more centrally located special industrial zones, all to be supported by heavy industry and especially by petrochemicals. Some twenty-five complexes (the previously mentioned *kombinato*), it was hoped, would encourage a proliferation of industrial growth and, as said, a possible future redistribution of the population. A crucial element in each site was electric power generation; hence first among facilities planned for construction in each location were ultramodern thermal power plants run by locally produced coal and imported petroleum. The sites of the new industrial agglomerations and of their satellite communities were all in — or were accessible to — coastal locations, allowing the direct importation of raw materials (including coal from other parts of Japan) for both manufacturing and power generation. Since much petroleum importation was anticipated, moreover, certain places required extensive port construction to permit the entry of increasingly large oil tankers, a manufacturing speciality of the Japanese shipbuilding industry. The formation of these centres progressed unevenly, with success, as stated, depending more or less on nearness to the core region, as obviously the need for labour and other personnel and the pull of the marketplace were powerful developmental factors. The most significant construction thus occurred where transportation was most accessible and where population centres were not too distant. Predictably, the Pacific coast between Kanto and Kyushu, including the Setonaikai region, was the area of most vigorous growth, with South Okayama (Okayama Kennan, or Mizushima) on Honshu, and the city of Sakaide in northeastern Shikoku, as leading examples. The more distant sites in such regions as Tohoku, southern Kyushu, and Hokkaido, found implementation much slower and more difficult, especially as the whole project assumed strong political implications, both locally and in Tokyo. None the less, in general there has been marked development as a result of this legislation, which is probably the most far-reaching of all the various postwar programmes. The human element played an important and perhaps unexpected role in complicating development plans in the 1970s because of the publication of former Prime Minister Kakuei Tanaka's book, *Building a New Japan, a plan for remodelling the Japanese Archipelago*, which, although the plan was not realised, had the effect of driving up land prices nationwide and ultimately of adding to inflationary tendencies.[3]

All these developments took place, however, before the interruption in petroleum supplies in 1973 and the subsequent derangement of the economy and when these occurred there was finally a major change in attitude toward planning. Now, statements that had always been a part of development programmes as to environmental protection or the maintenance of healthy living conditions, began to occupy a central place in projected schemes. Thus, the Land Use Planning Act of 1974 was couched distinctly in these terms. From this came the Third Comprehensive National Development Plan of 1978, which departed widely from the past by emphasising the creation of viable communities whose life would evolve from a combination of social as well as industrial and commercial activity. Regardless of language to the contrary, previous plans had all concentrated on the development and spread of industry, with much less regard for the comfort and well-being of the individual — a course which resulted mainly in exacerbating the very conditions the programmes were designed to ameliorate. Until the early 1970s, base populations and resulting environmental problems had multiplied to a point of near disaster in many places, especially in the already established urban-industrial core.

The new programmes were thus unprecedented in tone. Emphasis was now placed predominantly on essential human need and not, as before, on rapid economic growth with only token attention to possible consequences. The creation of some two to three hundred 'integrated residence areas' (*teijuken*) was called for, each of which would be a 'fundamental spatial unit for daily activity'.[4] These would roughly correspond to river basins and would have various features, both topographically and in terms of land use. Each would constitute a 'nodal area' surrounding a small or medium-sized city, whose ideal future population would lie between 250 and 500 thousand, and which would ultimately be capable of satisfying individual need for residence, work, and schooling, in that order.[5]

The foregoing are, of course, only plans, and it remains to be seen, in the face of changing and often unforeseen economic circumstances, whether these will become reality. On the other hand, when the provocation is severe enough, the Japanese record of acting on firm resolutions is impressive. The clearing of the atmosphere in the latter 1970s is an appropriate example. Observers who may have experienced the stifling air pollution caused mainly by industrial smoke and automobile exhausts in many Japanese cities before about 1975, cannot but note with approval the vast improvement in air quality after that time. As mentioned repeatedly, conditions of the environment had deteriorated so drastically between the late 1960s and early 1970s that the Government was forced to take strenuous action in passing such legislation as the Anti-Pollution Basic Measures Law of 1967, which was strengthened and toughened in later years, often in ways that might be considered extreme in other industrialised nations. Meanwhile, Japan fast became a world leader in waste contamination and anti-pollution research.[6]

## Planning and future prospects

As in the world as a whole, problems of planning in Japan have been subject to administrative complexities and to other frustrations that have

*Planning and future prospects*

8.4 The Ginza, Tokyo's famous central shopping street on a normal weekday afternoon. This is closed to automobiles on Sundays and holidays (see 7.4)

been felt on virtually every level of Government. Although most cities have been rebuilt since 1945, many plans are still being implemented while others have been interrupted or diverted by changing conditions. Successful execution by public agencies has been limited chiefly to transportation, to coastal housing and industrial enclaves, and, wherever basic plans have not been thwarted by such factors as population pressures, the demands of traffic, or by private construction, to parts of cities. Especially in Tokyo there has been in recent years major remodelling of whole subdistricts, and the international expositions — the 1964 Tokyo Olympiad, EXPO '70 in Kansai, the Winter Olympics of February 1972 in Sapporo, and, to some extent, EXPO '75 in Okinawa — have introduced, often with striking effect, a fresh spirit of enterprise, enthusiasm, and national consciousness. Private corporations, especially automotive, electrical, chemical, and textiles, have planned and built efficient, even attractive, communities; but as was mentioned in the previous chapter, the scale of these is usually small, and the surrounding areas may be in sharp contrast. The new university towns and the religious centres cited in Chapter 3 may also represent exceptions.

The subject of planning in Japan has been important enough to have attracted on a somewhat theoretical plane the attentions and energies of a number of private or semiprivate organisations. One of the most influential of these has been the Japan Center for Area Development Research, whose interests are global but whose activities have included frequent

## The Japanese landscape

high-level conferences of leading representatives of all field of planning in Japan, together with recognised international authorities, many of whom have long contributed, through the United Nations and similar agencies, to postwar planning and research in Japan. Other organisations whose efforts have offered valuable information and action in this regard, are the Tokyo Institute for Municipal Research, a coordinating and data-producing agency of distinguished lineage, and the Japan City Mayor's Association.

It has seemed appropriate to conclude this resumé of changing landscape patterns in Japan by directing the reader's attention to lines of future development. The forecast is admittedly more than a little discordant. It should be clear, however, that uncertainty resides mainly in the rapidly evolving urban-industrial landscape and only to a limited extent in the agrarian sector. What has been said, moreover, should not be construed as belittling the results achieved so far; it is intended, rather, to place the latter in a realistic perspective. Unfortunately, even when ambitious programmes are carried out, they are often followed by a rash of social problems which either were not or could not be foreseen; this is a worldwide phenomenon, even in areas where efforts to cope with such problems extend considerably further back in history than in Japan.

It is curious that, while there seem to be few precedents in the West worthy of Japan's emulation, Japanese planners are constantly on the

8.5  Mall in front of Sendai's new railway station, built in the late 1970s to accommodate trains of the new super-speed Tohoku *Shinkansen* line which is scheduled to open in the mid-1980s. Such construction is often a stimulus to the entire reconstruction and modernisation of urban areas

*Planning and future prospects*

lookout for models from other cultures, as is seen in their ceaseless organising of conferences with international 'experts'. All too often, in the opinion of this writer, these efforts have been focused on developments and experiences in the United States, when the concept of planning as an historical process that produces slower results by involving a broad segment of the community, illustrated to best advantage in parts of Northwestern Europe, would probably be more useful. By emphasising accomplishments in the United States, Japanese observers have looked to a geographical milieu that is for the most part entirely different from their own, and one to which man has therefore adapted himself far differently. Consequently Japanese observers may find it difficult to avoid being so impressed by the great sweep of construction, especially of gigantic highway networks with their swift flow of traffic under ideal conditions, that they overlook some of the untoward consequences, particularly in terms of immediate environmental quality and of the impact on the future life of the region — implications which may have been inadequately considered by the Americans when the plans were first drawn, and which, if circumstances now permitted, might have led to very different results. For example, Japan is presently in a stage of transportation development that seems to have certain parallels with that of the United States in the 1920s. Railroads are still the main means of travel and transport, and generally these are modern and efficient. On the other hand, fixed rail surface transport of all kinds in Japan is suffering increasingly stiff competition from mushrooming highway traffic, even to the extent of becoming a deficit

8.6 Crowd on sidewalk opposite Sendai station, August 1980

operation in many areas. Will the Japanese therefore, as did the Americans, allow the ultimate decline and death of these frequently excellent facilities, at least as passenger-carriers? Current action in the form of reductions in all but mainline services of the Japanese National Railways, or of the gradual abandonment of intra- and interurban rail systems in favour of buses and other automotive modes, would indicate that this may indeed be occurring. It is regrettable that the Japanese appear to be ignoring the warning implicit in Western, notably American, experience, and to be neglecting the lessons of those whose approaches, delayed perhaps by a slower pace of development, may have been more circumspect.

In the final analysis, however, this is no cause to assume an end to the characteristic Japanese technique of absorbing foreign influences in a selective fashion: the landscape that will eventually emerge will undoubtedly have a thoroughly modern air but an unmistakably Japanese flavour. Solutions to current problems will inevitably come from the Japanese themselves, and the genius for accommodation that has so marked the wonderful course of their evolution in the past will surely be the main factor in the Japan of the future.

# Notes

Publication details of books cited are given in the Bibliography.

## Chapter 1   The urban landscape

1. Linguists may differ about the origins of Japanese. Some feel these are still in doubt (Hattori Shiro in the *Encyclopedia Britannica*, Chicago, 1979 edn, vol. 10, p. 93); others see a distinct association with Ural-Altaic and Korean (R. A. Miller in the *Encyclopedia Americana*, New York, 1981 edn, vol. 15, p. 791).
2. See, for example, R. B. Hall, *Japan: industrial power of Asia*, 2nd edn, 1976, p. 24.
3. All Japanese plains are aggradational; see Ryuziro Isida, *Geography of Japan*, 1961, p. 22.
4. G. T. Trewartha, *Japan: a geography*, 1965, p. 22, and F. Takai et al, eds, *Geology of Japan*, 1963, p. 3, diagrammatically illustrate this divergence of opinion.
5. Trewartha, p. 22.
6. *Ibid.*
7. *Ibid*, pp. 35, 36.
8. *Nihon Kokuseizue* (A Charted Survey of Japan), 1979 edn, Table 6-1, p. 70.
9. *Ibid*, Table 23-3, p. 300.
10. *Ibid*, 1973 edn, Table 27-2, p. 268.
11. *Ibid*, 1979 edn, Table 33-2, p. 303.
12. *Ibid*, Fig. 33-1, p. 305.
13. *Ibid*, Table 25-6, p. 260.
14. *Ibid*, Table 25-1, p. 258.
15. For example, Table 18 of the volume, *Population of Japan: Summary of the Results of the 1975 Population Census of Japan*, pp. 274–97, shows that the nation's DIDs had their largest population segments in the 25–29-year age bracket and though this was also true of the nation's cities — but not of the nation's counties (*gun*) — the percentages recorded for DIDs were the highest of any of the categories shown. Table 16, pp. 268–9 of this source, also indicates that four of the ten largest cities (Tokyo, Kawasaki, Yokohama, and Nagoya) still had more men than women, the reverse of the average — albeit this

Notes

predominance for leading metropolises has tended to decrease since 1920 and 1925, when all these cities (other than Kita Kyushu which was not formed until 1962) had a majority of males.
16. See *Zenkoku Shi, Cho, Son Yōran 53* (National City, Town and Village Yearbook, 1978), p. 3.
17. Adapted from the *Statistical Yearbook of Japan, 1980*, Table 141, pp. 214–15.

## Chapter 2    The agrarian landscape

1. See: 'Agricultural policy outlook for the 1980s', *Japan Report*, 26, No. 2, 1 February 1980, p. 4.
2. Most of the material on the background to irrigated rice culture is based on the writings of the late geographer Ogasawara Yoshikatsu, especially his article 'The role of rice paddy [sic] development in Japan'. Other works, such as the fundamental book in English on the history of Japanese agriculture by T. C. Smith, *The Agrarian Origins of Modern Japan*, have been consulted.
3. This section incorporates the suggestions and comments of the well-known archeologist, Professor Richard J. Pearson. Prior mention of desirable qualities of wet rice other than its nutritional advantages refers partly to its ability to withstand long periods of storage without deterioration.
4. Descriptions of the socioeconomic organisation of feudal Japan and of the changes wrought by modernisation are, in part, from J. W. Hall and R. K. Beardsley, *Twelve Doors to Japan*, 1965, pp. 156–7.
5. R. P. Dore, 'Beyond the land reform: Japan's agricultural prospect', p. 273.
6. Tobata Seiichi, *An Introduction to Agriculture* [sic] *of Japan* (in English), 1958, pp. 48–9.
7. *Japan Report*, 9, 1 June 1980, p. 5.
8. *Ibid*, p. 4.

## Chapter 3    The city in Japanese history

1. Toyoda Takeshi, *Nihon no Hoken Toshi* (Feudal Cities of Japan), 1954, p. 38.
2. G. B. Sansom, *Japan: a short cultural history*, 1943, p. 356.
3. Kyoto became the capital in AD 794, succeeding Nara (although the town of Nagaoka, near by, was used for ten years during its construction), and while it remained the imperial seat until 1868, Tokygawa Ieyasu chose Edo (now Tokyo) as the headquarters of his régime, and from 1603 Kyoto was quickly eclipsed in size and importance.
4. Sansom, op. cit., p. 414.
5. Toyoda Takeshi and Orui Noboru, *Nihon Jokaku Shi* (Japanese Castle History), 1941, pp. 528–33.

6. Yazaki Takeo, *Social Change and the City in Japan*, 1968, p. 139.
7. *Ibid*, p. 133.

## Chapter 4    Changes in the urban landscape after 1868

1. Yazaki Takeo, *Social Change and the City in Japan*, 1968, pp. 478—85.
2. For a thoughtful account of the historical geography of Japanese highways, see R. B. Hall, 'The road in old Japan'.
3. J. K. Fairbank, E. O. Reischauer and A. M. Craig, *East Asia: the modern transformation*, 1965 edn, p. 247.
4. *Nihon Toshi Nenkan* (Japan Municipal Yearbook), vol. 10, 1941, p. 68. This figure was modified after 1924 to account for the amalgamation of surrounding towns, so that in this same reference work a more realistic figure of 85 133 for 1920 is also given (p. 60).
5. Ogasawara Yoshikatsu, *Nihon no Toshichiiki Bunkachiiki no Shihyo to Shite* (A Guide to Japanese Culturo-Urban Areas).

## Chapter 5    Historical aspects of the commercial landscape

1. G. B. Sansom, *Japan: a short cultural history*, 1943, p. 268.
2. *Ibid*, p. 354.
3. *Ibid*, p. 269.
4. H. Patterson Boger, *The Traditional Arts of Japan*, 1964, pp. 105—28.
5. E. O. Reischauer and J. K. Fairbank, *East Asia: the great tradition*, 1960, p. 563.
6. *Ibid*, p. 562, also see J. K. Fairbank, E. O. Reischauer and A. M. Craig, *East Asia: tradition and transformation*, 1973, pp. 381—4.
7. *East Asia: the great tradition*, p. 637.
8. *Ibid*, pp. 639—42, which covers the material given on Edo period commercial and industrial organisation.
9. C. D. Sheldon, *The Rise of the Merchant Class in Tokugawa Japan, 1600—1868*, 1958, p. 24.
10. *East Asia: the great tradition*, pp. 638—9.
11. *Ibid*, p. 641.

## Chapter 6    Landscapes of commerce and industry, 1868—1945

1. Among the advantages to Japan at that moment in history were the absence of real rivals in East Asia; the political and cultural unity of Japan; its long history of social stability, based on the Confucian ethic, which was staunchly maintained throughout the period of initial modernisation; and the country's extensive educational foundations, particularly a high degree of literacy which had allowed, long before 1868, the absorption by a small but vital contingent of *samurai* of many foreign innovations, not only in the sciences but in many other fields.

*Notes*

2. Much of this section on early industry is based on the excellent work of T. C. Smith, *Political Change and Industrial Development in Japan: Government enterprise, 1868–1880,* 1955 edn.
3. *Ibid,* pp. 5–6.
4. See for example, Seymour Broadbridge, *Industrial Dualism in Japan.*
5. The organisational framework of post-Meiji industrial expansion given here reflects that of a feature article in the Meiji Centennial Issue of the *Japan Times,* 23 October 1968.
6. J. K. Fairbank, E. O. Reischauer and A. M. Craig, *East Asia: the modern transformation,* p. 275.
7. *Japan Times,* 23 October 1968.
8. J. K. Fairbank, E. O. Reischauer and A. M. Craig, *East Asia: the modern transformation,* p. 260.
9. Many details of the economic development of Japan in the period covered in this chapter are based on the authoritative work of W. W. Lockwood, *The Economic Development of Japan, Growth and Structural Change, 1868–1938.*
10. P. H. Clyde and B. F. Beers, *The Far East,* 4th edn, 1966, p. 257.
11. Information on locational changes in industry is taken from Yoshio Okuda *et al,* 'Industrialisation and the growth of manufacturing cities in Japan', which neatly organises the complexities of this development, but more recent and detailed analyses are now available in Association of Japanese Geographers (ed.), *Geography of Japan,* especially in Chapter 12 by K. Murata, 'The Formation of Industrial Areas'; and Chapter 13 by Y. Miyakawa, 'The Location of Modern Industry'.
12. Lockwood, op. cit., p. 94.
13. *Ibid,* pp. 569–70.
14. *Ibid,* p. 66.
15. *Ibid,* p. 75.
16. Okuda *et al,* op. cit, pp. 58–60.
17. Fairbank *et al, East Asia: tradition and transformation,* 1973, p. 815.

## Chapter 7  Reconstruction and the growth of the contemporary industrial landscape

1. *Japan Times,* Meiji Centennial Issue No. 1, 23 October 1968, pp. 8–9.
2. Johannes Hirschmeier, 'The Japanese spirit of enterprise, 1868–1970', pp. 21–2.
3. *Japan Times,* Meiji Centennial Issue No. 1, 23 October 1968, p. 9.
4. It is characteristic of the salary structure of Japan to award bonuses at regular intervals, usually twice yearly. These payments, which are often equal to several months' salary, are used for major purchases, and customarily the remaining amounts are banked or invested – an important factor in general economic growth.
5. S. Broadbridge, *Industrial Dualism in Japan,* 1966, p. 12.

6. *Ibid*, pp. 93—4.
7. *Japan Report*, 17, No. 11, 1971, p. 6.
8. *Japan Times*, Meiji Centennial Issue No. 1, 23 October 1968, p. 9.
9. *Statistical Handbook of Japan, 1980*, p. 49 and *Nihon Kokuseizue*, 1979, Chart 29—3, p. 284.
10. *Statistical Handbook of Japan, 1980*, Table 24, p. 47.
11. *Ibid*.
12. *Ibid*, p. 57.
13. *Ibid*, pp. 44—5.
14. *Facts and Figures of Japan*, 1980, p. 72.
15. *Nihon Kokuseizue*, 1979, p. 133.
16. Okuda *et al*, in *Japanese Cities*, pp. 58—60.
17. Association of Japanese Geographers (ed.), *Geography of Japan*, 1980, Ch. 13 by Y. Miyakawa, 'The Location of Modern Industry', p. 296.
18. Okuda *et al*, op. cit., pp. 59—60.

## Chapter 8 The Japanese landscape

1. G. Sansom, *Japan: a short cultural history*, 1943, p. 465.
2. This section on planning is based mainly on the two final chapters of the book, *Geography of Japan*, Association of Japanese Geographers (ed.) (see footnote 11, Chapter 6). Chapter 17 by Kawabe Hiroshi is titled 'Internal Migration and the Population Distribution in Japan', and Chapter 18, by Kawashima Tetsuro, is 'The Regional Pattern of the Japanese Economy: its characteristics and trends'. Other materials that have provided important background information are: Honjo Masahiko, 'Trends in Japanese Development Planning', 20 January 1971, pp. 54—104; Kiuchi Shinzo, 'Recent reigonal development and planning in Japan', in *Festschrift Leopold Scheidl*, and Ito Tatsuo, 'A Review of urbanisation in Japan'.
3. Tanaka Kakuei, *Building a New Japan; a plan for remodelling the Japanese archipelago*. Material on the law for establishment of new industrial cities is partly from *Geography of Japan*, pp. 406—7.
4. *Ibid*, pp. 388, 410.
5. *Ibid*.
6. See 'New methods in the fight against pollution', *Japan Report*, 18, No. 6, 16 March 1972, 1—3.

# Bibliography

The works cited here are those used in the text only and the list is not meant to be exhaustive. On the other hand, many represent the most reliable sources available. Effort has been made to keep all references to a minimum and to avoid being parochial.

Japanese names are given as they would appear in Japanese: family names are consistently first and there is no comma in between.

Emphasis is obviously on materials that may be available in large libraries, in English, and the readers' indulgence is asked for the few unavoidable citations in the Japanese language.

## General works in English

CLYDE, PAUL H., and BEERS, BURTON F. *The Far East,* 4th edn, Prentice-Hall, Englewood Cliffs, N.J., 1966. One of several well-known texts on the political history of East Asia.

*FACTS AND FIGURES OF JAPAN 1980,* Foreign Press Center, Tokyo, 1979. A brief summary in English of statistical data published privately but distributed through consular offices and containing tables, graphs, and some text.

*JAPAN REPORT,* Consulate General of Japan, New York. A useful publication issued bimonthly in English and containing current information on Japan, arranged by topic. There is a yearly index.

*JAPAN STATISTICAL YEARBOOK,* Bureau of Statistics, Office of the Prime Minister, Tokyo. Excellent, official reference work in Japanese and English, published yearly by the Japanese government and showing basic data on many phases of activity, often with comparative tables, many of which pertain to individual cities.

*1975 POPULATION CENSUS OF JAPAN,* Bureau of Statistics, Office of the Prime Minister, Tokyo, 1976–80. The most recent quinquennial and complete census to be fully published and available in large libraries. It will be superseded by the 1980 Census, which is presently being analysed and issued as the most recent of the decennial censuses, dating from 1920. Since the Second World War, all censuses have been bilingual, while the Census of 1940 was republished in English and Japanese in 1960. Japanese population censuses have become increas-

*Bibliography*

ingly sophisticated and computerised, and give much detailed information on modern life, often compared to other years. Separate publications deal with marginal topics. For example, along with the censuses of 1960, 1965, 1970, and 1975, there are whole volumes dealing with densely inhabited districts (DID), and incorporating large-scale maps of all urban areas. There are, in addition, other more specialised censuses (e.g. of manufacturing, or of agriculture) published on a relatively regular basis by the government.

*STATISTICAL HANDBOOK OF JAPAN 1980*, Bureau of Statistics, Office of the Prime Minister, Tokyo, 1980. The most recently available issue of yearly summaries of statistical information, taken from Government sources and presented succintly in graphs and tables (often compared to other years), with commentary in English and, in other foreign editions, other languages. These helpful publications are usually available free at Japanese consular offices throughout the world.

*JAPAN TIMES*, Tokyo, various issues. One of several current English language daily newspapers published in Tokyo (the only one without a Japanese language counterpart). Contains foreign and domestic news, editorial opinion, feature articles, and a convenient summary of daily editorials from Japanese language newspapers. A reduced-size India-paper airmail edition is also published, and for libraries this is bound and equipped with a crude index.

## Books in English

ASSOCIATION OF JAPANESE GEOGRAPHERS (ed.) *Geography of Japan*, Teikoku-Shōin, Tokyo, 1980. A polished volume of articles on current themes in Japanese geography by twenty-six contemporary specialists. Similar in format to another publication edited by this association, *Japanese Cities: a geographical approach*, Association of Japanese Geographers, Tokyo, 1970, but larger and in more standard English.

BOGER, H. PATTERSON. *The Traditional Arts of Japan*, Doubleday, Garden City, New York, 1964. A representative work, nicely illustrated, on artistic expression, including such practical arts as sword-making, armour, ceramics, and leather craft. The text is scholarly, well-documented and deftly written.

BROADBRIDGE, SEYMOUR. *Industrial Dualism in Japan*, Aldine, Chicago, 1966. A rare volume in English on the important subject of traditional vs. modern industry, from the standpoint of an economist.

BURKS, ARDATH W. *Japan: profile of a postindustrial power*, Westview Press, Boulder, Colorado, 1981. A new and introspective view of modern Japan and its cultural and historical heritage, by a renowned American scholar.

CRESSEY, G. B. *Land of the 500 Million*, McGraw-Hill, New York, 1955. A standard work, now much dated, on the geography of China but

## Bibliography

with interesting and detailed information on physical background, and illustrated by excellent maps and other figures.

FAIRBANK, J. K., REISCHAUER, E. O., and CRAIG, A. M. *East Asia: tradition and transformation*, Houghton-Mifflin, Boston, 1973. A one-volume work which combines much of the material in the previous two-volume set, the second of which (*East Asia, the modern transformation*, 1965) is by the same authorities and publisher. The older work is more detailed in some respects but is perhaps not as current in the modern sections.

GINSBURG, N.S. (ed.). *The Pattern of Asia*, Prentice-Hall, Englewood Cliffs, New Jersey (1st edn. 1958). A long-recognised, authoritative, standard physical and cultural geography of Asia by a group of regional specialists.

HALL, R. B. *Japan: industrial power of Asia*, Van Nostrand, Princeton, N.J., 1976 (2nd edn). A highly useful summary of modern Japanese industrial growth and its impact on the world, especially on the United States.

HALL, J. W. and BEARDSLEY, R. K. *Twelve Doors to Japan*, McGraw-Hill, New York, 1965. A review of the history and culture of Japan in twelve chapters, mainly by the above authors. The condensation of a basic course on Japan then given at the University of Michigan.

ISIDA RYUZIRO. *Geography of Japan*, Kokusai Bunka Shinkokai, Tokyo, 1961. A small, well-illustrated volume in English by a noted Japanese geographer. The presentation is oriented on physical geography and, despite the absence of an index, is still a rich source.

KORNHAUSER, D. H. 'The role of geography and related factors in the rise of Japanese cities' (doctoral dissertation), University of Michigan, Ann Arbor, University Microfilms, 1956. Contains some of the basic information for this volume, especially for Chapter 3.

LOCKWOOD, W. W. *The Economic Development of Japan, Growth and Structural Change, 1868–1938*, Princeton University Press, Princeton, 1954. Long-recognised and highly regarded account of modern Japanese economic development, from Meiji times to the Second World War. A standard reference work by an important authority.

REISCHAUER, E. O. and FAIRBANK, J. K. *East Asia: the great tradition*, Houghton-Mifflin, Boston, 1960. The first of the two-volume set (mentioned above) on the history of East Asia, covering the time before major modernisation. Despite the enormous sweep of the book, it is authoritative, detailed and valuable as a reference tool.

SANSOM, G. B. *Japan: a short cultural history*, Appleton-Century, Crofts, New York, 1943. A classical history of the Japanese before the age of writing to the time of Tokugawa decline. This delightful volume, written with a sympathetic wit and deep understanding by a towering figure in the field, provides valuable insights into the culture.

SHELDON, C. D. *The Rise of the Merchant Class in Tokugawa Japan, 1600–1868: an introductory survey*, Association for Asian Studies,

New York, Monograph 5, 1958. An excellent, well-documented review of the subject by a dedicated modern scholar.

SMITH, T. C. *The Agrarian Origins of Modern Japan*, Stanford University Press, Stanford (1st edn 1959). An important, fundamental reference work, skilfully presented by a foremost authority on Japanese economic history.

SMITH, T. C. *Political Change and Industrial Development in Japan: Government enterprise, 1868—1880*, Stanford University Press, Stanford (1st edn 1955). A fully analytical account of early Meiji period industrialisation.

TAKAI F., MATSUMOTO T., and TORIYAMA R. (eds). *Geology of Japan*, Tokyo University Press, Tokyo, 1963. An interesting and rare work published in Japan in English on the foundations of Japanese geology.

TANAKA KAKUEI. *Building a New Japan: a plan for remodeling the Japanese Archipelago*, Simul Press, Tokyo (1st edn 1972). A controversial work by a controversial figure in modern Japanese politics. Translated.

TOBATA SEIICHI. *An Introduction to Agriculture of Japan [sic]*, Agriculture, Forestry and Fisheries Productivity Conference, Tokyo, 1958. A good summary of agricultural developments in the postwar era, up to then.

TREWARTHA, G. T. *Japan: a geography*, University of Wisconsin Press, Madison, 1965. A prominent regional geography of Japan with excellent maps and other illustrations. Originally called *Japan: a physical, cultural, and regional geography* (1945), although the new edition is much enlarged and updated.

YAZAKI TAKEO. *Social Change and the City in Japan*, Japan Publications Trading Company, Inc., Tokyo, 1968. An unusually detailed work in English stressing sociological aspects of Japanese urbanisation from early history to about 1920. Translated.

## Articles in English

DORE, R. P. 'Beyond the land reform: Japan's agricultural prospect', *Pacific Affairs*, Vol. 36, 1976, pp. 265—76. A thoughtful summary of the situation in 1963 by a noted Japan specialist and authority on land reform, its background and problems.

DRUCKER, P. F. 'Japan: problems of success', *Manchester Guardian Weekly*, 28 May 1978, pp. 1, 8; 4 June 1978, p. 9. Analysis of the rapid ageing of the population by a leading economist with special interest in Japan.

HIRSCHMEIER, JOHANNES. 'The Japanese spirit of enterprise, 1868—1970', *Business History Review*, Vol. 44, No. 1, Harvard University Press, Boston, 1970, pp. 13—38. Presents valuable insights into the inner workings of the Japanese business world since the Meiji period.

## Bibliography

The volume contains much enlightening information on various aspects of Japanese business.

ITO TATSUO. 'A review of urbanisation in Japan', in Association of Japanese Geographers (eds), *Japanese Cities: a geographical approach*, Tokyo, 1970, pp. 257—64. A brief but thorough summary of urban development in Japan by a leading English-speaking Japanese geographer.

HALL, R. B. 'The road in old Japan', in *Studies in the History of Culture*, American Council of Learned Societies, New York, 1942, pp. 122—55. A pioneer work in English by a respected authority on Japanese geography.

KAWABE HIROSHI. 'Internal migration and the population distribution in Japan', in Association of Japanese Geographers (eds), *Geography of Japan*, Teikoku-Shoin, Tokyo, 1980, pp. 379—89. A recent summary of contemporary and projected plans for the remainder of this century is presented in the last portion of this article, pp. 387—9.

KAWASHIMA TETSURO. 'The regional pattern of the Japanese economy: its characteristics and trends', in Association of Japanese Geographers (eds), *Geography of Japan*, Teikoku-Shōin, Tokyo, 1980, pp. 390—414, with particular attention to planning and its philosophy, pp. 409—14.

KIUCHI SHINZO. 'Recent regional development and planning in Japan', in *Festschrift Leopold G. Scheidl:Zum 60. Geburtstag*, Wiener Geographisches Schriften, Nr. 24—9, Part II, Vienna, 1967, pp. 63—5. A valuable review of the subject, as of the time, by a leading Japanese urban geographer.

MURATA KIYOJI. 'The formation of industrial areas', in Association of Japanese Geographers (eds), *Geography of Japan*, Teikoku-Shōin, Tokyo, 1980, pp. 249—64. A modern industrial geographer's view of the whole matter of regionalism in industry in contemporary Japan.

MIYAKAWA YASUO. 'The location of modern industry in Japan', in Association of Japanese Geographers (eds), *Geography of Japan*, Tokyo, 1980, pp. 265—98. Similar to the above in scope but with detailed emphasis on specific industries and industrial growth.

OGASAWARA YOSHIKATSU. 'The role of rice paddy [sic] development in Japan', *Bulletin of the Geographical Survey Institute*, Tokyo, No. 5, Part 4, March 1958. A detailed presentation of the evolution of paddy culture in Japan, from prehistoric times to the 1950s.

OKUDA YOSHIO, OTA ISAMU, TAKAHASHI NOBUO, and YAMAMOTO SHIGERU. 'Industrialisation and the growth of manufacturing cities in Japan', in Association of Japanese Geographers (eds), *Japanese Cities: a geographical approach*, Association of Japanese Geographers, Tokyo, 1970, pp. 53—62. Another authoritative article in this work containing special information on landscape change at that time, and as demonstrated in Chapters 6 and 7 of the present volume.

# Bibliography

## General works in Japanese

*NIHON KOKUSEIZUE* (A Charted Survey of Japan), The Tsuneta Yano Memorial Society, published by Kokusei-sha, Tokyo, various dates, the latest here being 1980. A privately published yearly statistical volume compiled from official sources, with commentary and illustrations on many aspects of Japanese life, often showing comparisons with previous data. This reliable work is often seen in academic circles regardless of the discipline. It is paralleled by a good English edition (*Nippon, A Charted Survey*), available widely in libraries.

*NIHON TOSHI NENKAN* (Japan Municipal Yearbook), 1941 and other dates, Tokyo Institute for Municipal Research. The final pre-war issue in a distinguished series of statistical publications dealing entirely with cities. These were compiled with great care and published from the early 1930s until 1953 by this organisation, and this particular issue contains rare urban data on population in the Meiji period, an unusual feature. After 1945 the publisher was changed from the Tokyo Institute for Municipal Research to the Japan City Mayor's Association...

OGASAWARA YOSHIKATSU. 'Nihon no Toshichiiki Bunkachiiki no Shihyo to Shite' (A guide to Japanese Culturo-Urban Areas), *Sundai Shigaku (Sundai Historical Journal)*, No. 4, Special Issue, March 25, 1955, pp. 107–30, maps.

SATO H. and NISHIKAWA O. *Shinsho Koto Chizu* (A New and Detailed Senior Atlas), Teikoku-Shōin, Tokyo, 1977. A recent example of a selection of fine atlases for general use in Japan. Used here for certain details and specific information on such as climate and other physical matters.

*SHIN SANGYO TOSHI NO GENKYO* (The Present State of the New Industrial Cities), Economic Planning Agency, Tokyo, 1967. A descriptive pamphlet from official sources on the development of the plan for the construction of new industrial cities and special industrial zones (see maps, Chapter 8).

*TOSHI MONDAI* (Municipal Problems), Japan Institute for Municipal Research, various dates. A monthly periodical on urban problems containing extensive bibliographical information in Japanese and other languages.

TOYODA TAKESHI. *Nihon no Hoken Toshi* (Feudal Cities of Japan), Iwanami Shoten, Tokyo (1st edn), 1952. A leading historical work.

TOYODA TAKESHI and ŌRUI NOBORU. *Nihon Jōkaku Shi* (Japanese Castle History), Yuzankaku, Tokyo (1st edn), 1936. An important text by two renowned authorities.

*ZENKOKU SHI, CHŌ, SON YŌRAN 53* (National City, Town, and Village Yearbook 1978), Dai-Ichi Hōki, Tokyo, 1978. The most recently available issue of a yearly publication devoted to many details of all cities, though it does not cover DID data. Published under government auspices.

# Index

Absentee landlords, 45, 46, 48. 147
Abukuma Highlands, 6 (Map 2), 7
*Accommodation* (Japanese cultural attribute), 1, 2, 113, 144, 162
Administration, 32–3, 61, 64, 66, 71–2, 76, 86, 92, 158
Africa, 118
Agrarian landscapes, 34–57, 120, 124, 145–8
  modern, 46–54, 120, 124
  traditional, 21–2, 34, 40–2, 120, 124, 145–8
Agrarian poverty, 23, 45, 46, 47, 49, 50, 54, 56, 112, 116, 118, 124, 146–7, 151
Agriculture
  cooperatives, 50–1
  development, 37–53, 55–7, 99, 105, 145–8
  implements, 37, 52, 53, 56
Agricultural Land Law (1952), 48, 52, 58
  *See also* Land reform
Agricultural mechanisation, 35, 39 (Fig. 2.2), 42, 44 (Fig. 2.3), 45, 50–3, 51 (Fig. 2.5), 55 (Fig. 2.6), 56
Agricultural problems
  of the past, 42–3, 99, 112, 116, 119, 147
  present, 45–57, 147
Agricultural products and trade, 50–3, 55–6, 67, 99, 116, 136, 146
Aichi (*Ken*), 31 (Map 11), 37, 68–9 (Table 3.1), 75, 115, 140
Ainu, 59
Aircraft industry, 116, 124
Airways, 16, 18 (Map 9), 92
Akita
  *Ken*, 17, 31 (Map 11), 51, 68–9 (Table 3.1)
  *Shi*, 17, 68–9 (Table 3.1), 147
Akita Plain, 41

Alluvial
  fans, 38, 40
  lowlands, plains, 10–15, 40, 55, 59, 66–7, 146, 163 (Fn. 3)
Amagasaki (*Shi*), 120, 139
Amakusa island group, 5
Amalgamation (urban), 30
Animal husbandry, 51–2, 146
Anti-Pollution Basic Measures Law (1967), 158
Aomori (*Ken*), 31 (Map 11), 68–9 (Table 3.1)
Aomori (*Shi*), 109
Arcuateness, 4–8, 8 (Map 3), 12
Area (size), 4–5, 15–17
Ariake Sea, 41, 42, 139
Army (imperial), 45
Artistic expression, 59–60, 94–6, 136, 144–5, 150
  aesthetics, 20, 21, 59–60, 95–6, 136
  applied arts, 20–1, 59, 73, 75, 95–6
  in sculpture, 21
Asahikawa (*Shi*), 78
Asia, 5, 49, 75, 94, 104, 114, 117, 118, 151, 154
*Ashikaga* period (AD 1336–1568), 62
Ashikaga (*Shi*) (textile city, Tochigi *Ken*), 138
Aso (volcanic, caldera crater), 8, 9 (Fig. 1.2)
Atami (*Shi*), 74, 88
Atmospheric cycle, 12, 13 (Map 6)
Automobiles, 16, 135, 138, 140, 150, 152 (Fig. 8.3)
Automotive industry and rolling stock, 85, 116, 129, 130, 131, 135, 138, 140, 143, 161
Awaji Island, 5

'Baby-boom' (1948), 23
Backmarshes (early paddy lands), 38

175

## Index

*Baiu* (monsoonal rainy season), 36
Barter system (of rice exchange), 40
Bed towns, 83, 120, 124
 *See also* Satellite cities; Dormitory towns
Beer brewing, 85, 136
Beppu (*Shi*), 74, 88
Birth rate, 23
Biwa-*ko* (Lake), 59, 75, 78
Black (or Japan) current (*Kuroshio*), 19
'Blue-collar towns', 120
Bombing (by Allied aircraft, 1945), 119, 121–2, 124, 154
Bonin Island group, 4
*Bonsai* (dwarfed tree culture), 144
British
 influence in railway construction, 79–80
 influence in pre-Meiji period industry, 111
 as purchasers of spinning equipment, 115
 as shipbuilding rival, 135
Buddhism, 22, 58–9, 61, 74, 88, 96–7, 105, 144
 *See also* Zen
Bungo Channel and Strait, 5 (Fig. 1.1), 9
Buses, 16, 90, 162
By-pass routes (*baipassu*), 82

Cadastral (rectilinear) survey, 38
California, 5
Canals, 40, 81
Carpentry and woodworking, 75, 102
Castle town (*jōkamachi*), 64–72, 68–9 (Table 3.1), 66–7, 63 (Map 12), 70–2, 75–6, 84–7, 91, 92, 98, 148–9
 castle structures, 40, 66, 98, 150
 as modern prefectural capitals, 68–9 (Table 3.1), 66, 77, 78, 84, 87, 91, 92, 121, 153
 site characteristics before and during Edo period, 66–70, 76, 78, 148
*Ceiling* rivers, 70
Cement, 19, 125, 136
Central Business Districts (CBD) of modern cities, 83–4, 86, 92
Ch'ang An, 60
Chemicals industry, 116, 120, 124, 125, 135–6, 138–9, 157
Chiba (*Ken*), 31 (Map 11), 33
Chiba Peninsula, 19

Chiba (*Shi*), 139
China, 1, 22, 37, 56, 58–61, 75, 94, 97, 98, 105, 111, 114, 115, 117, 118, 123, 131, 144–5, 148, 153
Chinese philosophy and religion, 58–9
*chō* (town), 32, 103
Choshu (*Han* of southwestern Honshu), 108
Christianity, 98, 105
 *See also* Religion
Chubu (region made up of various prefectures and sub-divided into Hokuriku, Tosan, and Tokai, 26 (Table 1.4), 31 (Map 11), 155
Chubu (Mt.) Node, 6
Chubu Region Development Law (1966), 155
Chugoku (region made up of various prefectures of western Honshu), 17, 26 (Table 1.4), 31 (Map 11), 125, 141
Chugoku Highlands, 6
Chukyo (Nagoya) Region, 41, 138
Chuo line (Japanese National Railways), 81, 117, 120
Cities, 14–15, 47–8, 54–6, 58–76, 77–93, 121, 123, 137, 144, 148–50, 153–61
 *See also other headings*
City (*Shi*), 26, 30 (Table 1.8)
City planning, 59–60. 153–8
 *See also* Metropolitan; Planning; Urban, etc.
Civic centres (of modern prefectural capitals), 85
Climate, 10–15, 36, 38, 45, 53, 55
Climatic regions, 17
Clothing, 85, 125
 *See also* Textiles
Coal, 15, 17, 111, 115, 120, 127, 133, 137
Coasts, 8–10, 115, 125, 138–9, 143
Cold war, 130, 142
Communications, 18 (Map 9), 17, 47, 61, 85, 91–2, 123, 148
Communist Party, 50
Commuting, 29, 56, 84, 86, 120, 127, 142
 *See also* 'Part-time farming'
Compost, 37
Comprehensive National Land Development Act (1950), 155
Confucianism, 59, 119, 144
 *See also* Philosophy

176

# Index

Constitution, postwar, 130
Construction industry, 15, 21, 127, 136, 139–40, 151, 159
Consumer goods, 117, 122, 128–31, 133
Copper, 17, 95, 120
 See also Mining (copper)
Copper 'cash', 94, 97, 105
'Core' (urban, or industrial), 4, 15, 16 (Map 8), 17, 19, 22, 25, 78, 80, 81, 86, 99, 103, 108, 118, 137, 141, 150, 157–8
Cotton spinning, 75, 115, 123, 132
 See also Textiles (cotton)
Crops, 36, 51–3, 56
Cryptomeria (giant Asian conifers), 21
Cultural identity, 1, 22, 94
Cultural regionalism, 17, 71, 72
Culture (Japanese), 58, 94, 112, 119, 123–4, 127, 130–1, 140, 145–8, 150, 152, 153, 162
Cyclonic storms, 12, 36

Daimyō (fief [Han] administrators), 42, 56, 66–7, 97–8, 103, 111, 122, 151
Dairying, 37, 51
Danchi (housing [and other] enclaves or estates), 90 (Fig. 4.5), 90, 138
 See also New Town
Dazaifu (Heian period provincial capital and planned city, northern Kyushu), 60
'de facto capitals', 61
 See also Kamakura
Death rate, 23
Deconcentration, Educational, 85
Defence cities, 78, 90, 93
Deltas and deltaic lands, 6, 41, 70
Demilitarisation, postwar, 130
Democracy, 107, 112
 in the 1920s, 117
 since 1945, 127
Densely inhabited districts (DID), 27 (Table 1.5a and 1.5b)
Depression (economic, 1929–39), 45, 151
Dewa (Mt.) range, 6
Dialects, 71–2
Diet (Japanese Parliament), 48
Dikes, 41
 See also Waju
Diluvial lands, 42, 45, 50, 70
Dō (Hokkaido) (Island Prefecture), 32
Dogo (onsen, or spa) of Matsuyama (Shi), 89
Dormitory towns, 83, 120, 124
 See also Bed towns; Satellite cities
Drainage, 12
Dry field crops, 37, 52
Dual industrial structure, 99, 112, 131–4, 142
Dual nucleae (of cities, caused by railway construction, 19th century), 83–4, 92
Dutch influence (Edo period), 98, 108–9

Earth movement, 5–8, 11
East Asian monsoon, 12–13, 13 (Map 6), 36, 104
 See also Monsoon
Echigo (Niigata) Plain, 14, 38, 40
Economy, 25, 32–3, 40–6, 46–50, 54, 56, 58, 64, 67, 76, 94, 97, 99–104, 105–6, 107, 111–14, 116–18, 123–4, 126–9, 131–5, 138, 142–3, 145–52, 154, 156–8
Edo (modern Tokyo), 32, 40, 42, 64, 67–71, 68–9 (Table 3.1), 76, 78, 108, 148
Edo (Tokugawa) period (1600–1868), 22, 40–2, 56, 64–74, 76, 77–9, 87, 91, 94–106, 107–11, 145, 148–9, 153
Edokko ('child of Edo'), 72
Education, 23, 85, 91–2, 112–13, 123, 159
Educational cities, 85, 91, 92, 159
Ehime (Ken), 31 (Map 11), 33, 68–9 (Table 3.1)
Equalisation grants (for cities), 33
Electric power, 17, 21, 114, 116, 126 (Fig. 7.1), 127, 133–4, 136–7, 143
 hydro, 21, 116, 133
 nuclear, 133, 143
 thermal, 17, 137, 143
Electrical industry (electronics), 120, 132–5, 153
 See also Precision instruments
Electrification, 17, 80–1, 84, 89–90, 124
 See also under Railways
Environmental control (agrarian), 36–7, 40–3, 52, 55, 144–5
Environmental pollution and public nuisance, 20, 54, 81, 138, 140–1, 142–3, 153, 157–8, 161–2

177

*Index*

See also Kogai
Ethnicity (Japanese), 22
Europe, 45, 58, 61, 67, 76, 98, 101–2,
    107, 112, 114, 118, 129–30,
    135, 137, 149, 156
Exclusion (Edo period), 62, 65, 98,
    102 fn., 105
EXPO '70, 89, 159
    See also Senri New Town
EXPO '75 (Okinawa), 159
Export trade
    Pre-Edo, 94–6, 97, 105
    Edo period, 97–101, 114–15,
        151–2
    Meiji period, 101, 114–15
    post-Meiji, 115–18, 123–4, 151–2
    after 1945, 128, 130, 134–7

Family and clan, 23, 37, 47, 54, 57,
    59, 61, 67, 95–6, 118–19,
    151
Farmers, 41–3, 45–6, 51, 55, 74,
    146–8
    See also Horticulturalists; Social
        classes
Fault lines, 6
Fertiliser Industry and Production,
    37, 52, 54, 56, 114, 125,
    127, 154
Feudalism, 40–2, 61–2, 66–7, 77, 79,
    86, 91, 94, 96, 107, 112,
    122, 145, 149
Field consolidation, 37, 54, 57, 147
    See also Fragmentation; Land-
        holdings
Finance, banking and stock market, 131
First World War see World War I
Fishing, 19–21, 72, 52, 87, 146
Flora and fauna, 17, 21–2
Folk art, 95
Food, 19–21, 37, 47–8, 51–4, 99, 102,
    104, 119, 122, 136, 141,
    146, 150–51
    rationing (World War II), 47, 119
    shortages (postwar years), 47–8,
        122
Food and related industries, 51–2, 75,
    85, 102, 104, 136, 150
Foreign influences and advisers, 34, 79,
    98, 108–11, 122–3, 133,
    151, 161–2
Forestry, 21–2
Forests, 21–2, 37
*Fossa Magna*, 6, 7 (Map 2)
France, 15, 109, 135
Fragmentation, 37, 54, 57, 147

See also Field consolidation; Land-
    holdings
'Free' cities, 61–3, 72, 75–6, 89
    See also Hakata; Sakai
*Fu* (urban prefecture, Osaka and Kyoto),
    32–3
Fuji, 6
'*Fukoku kyohei*' ('national wealth and
    military strength') Meiji
    period slogan, 106
Fukui (*Ken*), 31 (Map 11), 137
Fukuoka (*Ken*), 31 (Map 11), 68–9
    (Table 3.1)
Fukuoka (*Shi*), 15, 63, 68–9 (Table
    3.1), 80, 85
Fukushima (*Ken*), 31 (Map 11), 33,
    68–9 (Table 3.1)
Fukushima (*Shi*), 68–9 (Table 3.1)

Gas and electric facilities, 114, 154
*Genroku era* (Edo period, 1688–1704),
    103
Germany and German influences, 129,
    135, 156
Gifu (*Ken*), 31 (Map 11), 33, 68–9
    (Table 3.1)
Gifu (Mt.) Node, 6, 8 (Map 3)
    See also Chubu Node
*Gokaido* (five main feudal highways),
    79
    See also Highways; Roads; trans-
        portation (land)
Government sponsorship of industry,
    111, 113–14, 116, 122–3,
    139, 143, 152, 157–8
Grasslands, 37
Great Britain, 9, 79–80, 111, 115, 135
    See also British, English, England
Green manure, 37
Gross National Product (GNP), 119,
    152
Guild organisations (*Za*) (Edo period),
    95–7, 150–1
    See also Monopolies
Gumma (*Ken*), 31 (Map 11), 33, 68–9
    (Table 3.1), 104

Hachiro Lagoon (reclaimed lake, Akita
    *Ken*), 51, 147
Hagi (*Shi*), 63 (Map 12), 68–9 (Table
    3.1)
Hakata, 62, 63 (Map 12), 97
    See also 'Free' cities
Hakodate (*Shi*), 78, 86
Hamamatsu (*Shi*), 88
*Han* (feudal fiefs), 56, 66–7, 99, 103,

178

*Index*

108, 111
Handicrafts, 89, 94—5, 99, 150
  *See also* Manufacturing, ancient
Hankyu Railway company, 89
Hanshin (Osaka-Kobe region), 32
  *See also* Kansai
Hanshin metropolitan area plan (1961—62), 155
Harbours, 9, 61—2
Heian period (AD 784—1184), 38, 58, 60—2, 75, 78, 94—5
Heiankyō, 60
  *See also* Kyoto
Heijokyo, 59
  *See also* Nara
Heike Story (*Heike Monogatari*, historical novel), 61
Hideyoshi (Toyotomi), 41, 64, 65 (Fig. 3.2), 66, 97
Highways, 16, 30—1, 67—70, 76, 78—9, 81—2, 88, 92, 149—50, 161—2
  Edo period, 30—1, 63 (Map 12), 67, 70, 76, 78—9, 88, 92, 149; *see also* Gokaidō
  modern, before World War II, 47, 78—9, 148
  since the 1950s, 47, 78, 79—82, 82 (Figs. 4.1, 4.2), 88, 91—2, 100—1 (Figs. 5.1, 5.2), 104 (Fig. 5.4), 121 (Fig. 6.4), 126—7, 146 (Fig. 8.1), 152 (Fig. 8.3), 161—2
  *See also* Roads; Transportation (land)
Hikone (*Shi*), 68—9 (Table 3.1)
Himeji (*Shi*), 68—9 (Table 3.1), 86, 139
Hirosaki (*Shi*), 68—9 (Table 3.1)
Hiroshige (woodblock print artists of the Edo period), 73
Hiroshima (*Ken*), 15, 31 (Map 11), 33, 68—9 (Table 3.1), 125, 128 (Fig. 7.2), 139 (Fig. 7.3)
Hiroshima (*Shi*), 68—9 (Table 3.1), 71, 121
History, 37, 58, 78, 94—5, 122, 145
Hitachi (*Shi*) and electric company, 17, 120, 133
Hizen (*Han*; now Saga *Ken*), 107, 110—11
Hokkaido, 2, 4, 6—8, 26 (Table 1.4), 31 (Map 11), 42, 45, 49—50, 56, 78, 85, 86, 90, 91, 113, 115, 122, 123, 125—6, 138, 141—2, 149, 152 (Fig. 8.3), 159

Hokkaido (Mt.) Node, 4, 6—7, 8 (Map 3)
Hokuriku district (part of Chubu), 31 (Map 11)
Home improvements, 46 (Fig. 2.4), 47
Homma (Yamagata *samurai* family), 41
Honda Motors Corporation, 133
Hong Kong and Singapore, 131
Honshu, 4, 6, 9, 12—19, 26 (Table 1.4), 37—8, 42, 59, 71, 80, 108—9, 115, 123, 125, 127—8, 138, 143
Horticulturalists, 147—8
  *See also* Farmers
Hot springs (*onsen*), 74, 76, 88—9, 93
Housing, 15, 23, 25 (Fig. 1.4), 54, 90, 93, 109 (Fig. 6.1), 110 (Fig. 6.2), 120, 124, 138, 153
Human Resources, 22, 102, 107—8, 145—6, 158
Hyogo (*Ken*) (also ancient port), 31 (Map 11), 33, 68—9 (Table 3.1)

Ibaraki (*Ken*), 31 (Map 11), 33, 68—9 (Table 3.1), 108
Ikeda Hayato (former Prime Minister and author of the 1962 Income Doubling Plan), 134, 156
Imabari (*Shi*) (textile city, northern Shikoku), 125
Image
  of cities, 32—3
  of Japan and the Japanese, 115—16, 118—19, 124, 130, 141—2, 153
Imperial Rescript on Education (1890), 112
Imperial universities, 85
Import trade, 94, 97, 104—5, 112—13, 115—19, 134—7
Income, 29, 54, 56, 66—7, 89, 134, 156
'Incipient capitalism', 107
India and Pakistan, 118, 131
Individualism, 145
Industrial cities, 83, 93, 115—16, 120—1, 123—4, 125—9, 137—41, 141, 143
Industrial diversification, 117, 120
Industrial 'estates' or 'parks', 138—9
Industrial growth 'poles' (1962), 156—8
Industrial modernisation, 42, 85, 98—9, 107, 108—24, 125—43, 150—3
Industrial rationalisation, 113, 117,

179

## Index

124, 132
Industrial revolution, 1, 98–9, 102, 106, 112, 151
Industries
  capital intensive, 114–15, 122–4, 132–3, 134, 142, 145
  labour intensive, 95–7, 115, 142, 145
  modern industries, 95, 107–24, 125–43
  traditional industries, 85, 94–7, 99–100, 102–4, 106, 131–2
  *See also various headings*
Industry, 15, 26, 46, 95–7, 101–4, 104–5, 107–24, 125–43, 150–3, 156–8, 160
Infanticide (*mabiki*), 23
Inland Sea *see* Setonaikai
'Inner' and 'outer' (tectonic) zones of 'old Japan', 6, 7 (Map 2)
Integrated Residence Areas (*teijuken*), 158
Inter-urban railways, 80, 84, 91–2
  *See also* Japanese National Railways (JNR); Private railways; Underground railways
Iron and steel industry, 17, 75, 108–9, 111, 114–16, 127, 134–5, 139, 142
Irrigated rice, 34–54, 59, 99, 144–8
  *See also* Paddy; Rice
Irrigation and drainage, 11–15, 21, 37–40. 45, 47
Ise Bay, 9, 41, 138
Ishikari Plain (Ishikari-Yufutsu Plain), 6, 13
Ishikari River, 6
Ishikawa (*Ken*), 31 (Map 11), 68–9 (Table 3.1), 109
Italy, 135
Ito (*Shi*), 74
Iwakuni (*Shi*) (airbase of western Honshu), 90
Iwate (*Ken*), 31 (Map 11), 68–9 (Table 3.1), 115, 127
Izu Peninsula, 6

Japan Center for Area Development Research, 159
Japan City Mayor's Association, 160
Japan (or Black) current (*Kuroshio*), 19
Japan Optical Corporation (Nikon), 129
Japan Sea *see* Sea of Japan
*Japan Times,* 122

Japan Trench, or Deep, 5
Japanese language, 1, 21, 58–9, 163 (n. 1)
Japanese National Railways (JNR), 80–1, 89, 91–2, 120, 162
  *See also* Railways
*Jō-ri* system (cadastral survey), 38, 60, 75, 153
  *See also* Rectangular grid
Joint enterprise (Japanese-foreign), 128, 134, 142

*Kabu-nakama* (Edo period commercial associations), 100
Kagawa (*Ken*), 31 (Map 11), 33, 68–9 (Table 3.1)
Kagoshima (*Ken* and *Shi*), 31 (Map 11), 68–9 (Table 3.1), 71, 108, 111
Kamaishi (*Shi*) (iron-mining and manufacturing city, Iwate *Ken*), 115
Kamakura (*Shi*), 61, 78, 153
  *See also* 'de facto' capitals'
Kanagawa (*Ken*), 31 (Map 11), 33, 68–9 (Table 3.1)
Kanazawa (*Shi*), 68–9 (Table 3.1), 71, 121
Kanoya (*Shi*) (airbase), 90
Kansai (Osaka-Kobe region), 32, 84, 124, 140, 150
  *See also* Kinki; Kinai; Hanshin; Keihanshin
Kanto (district), 26 (Table 1.4), 31 (Map 11), 37, 42, 61, 74, 83–4, 104, 109, 115, 120, 124, 149–50. 150–1, 153, 157
Kanto earthquake (1923), 153
  *See also* Earth movement
Kanto tectonic line, 7 (Map 2)
Kanto (Tokyo) Plain, 13, 14, 104, 109, 138
Karafuto (Sakhalin), 88, 120, 139
Kawasaki (*Shi*), 83, 88, 120, 139, 165 (Chap. 4, n. 4)
Keihanshin (Kyoto-Osaka-Kobe), 32
  *See also* Hanshin; Kansai, Kinki; Kinai
Keihin (Tokyo-Yokohama), 32, 140
*Ken* (Prefecture), 32–3
Kii Channel, 9
Kii Peninsula, 6
Kinai (old term for Osaka-Kyoto-Nara-Kobe region), 32, 37–8, 40–1, 78

*See also* Hanshin; Kansai; Keihanshin, Kinki
Kinki (Osaka-Kobe-Nara-Kyoto region), 26 (Table 1.4), 31 (Map 11), 32, 155
*See also* Hanshin; Kansai, Keihanshin, Kinai
Kinki Region Development Law (1965), 155
Kishiwada (*Shi*), 68–9 (Table 3.1)
Kitakami Highlands, 6
*See also* Abukuma Highlands
Kita Kyushu (*Shi*) (city incorporated 1963, amalgamating former cities of Moji, Kokura, Yawata, Tobata and Wakamatsu, which are now wards), 140
Kobe (*Shi*), 15, 33, 77, 79, 84, 92, 102, 115, 120, 139, 153, 155
*See also* Outports, Port City
Kochi (*Ken* and *Shi*), 31 (Map 11), 68–9 (Table 3.1)
Kofu Basin, 38, 52
Kofu (*Shi*), 68–9 (Table 3.1), 71
*Kofun* (pre-historic tomb mounds), 59
*Kōgai*, 134, 138, 140–1, 143, 150, 158
*See also* Environmental pollution; Public nuisance
Kojima Bay and Peninsula (reclaimed area, south Okayama), 41–2, 51, 125
*Kokudaka* (Edo period rice allotments), 67, 71
Kokura (former *Shi*, now *Ku* of Kita Kyushu), 68–9 (Table 3.1)
*See also* Kita Kyushu
*Kombinato* (industrial 'estates', 'parks' or enclaves), 134, 138–9, 157
Korea, 37–8, 56, 58–9, 99, 105, 114, 123, 145, 163 (n. 1)
Korean War (1950–53), 130–1, 133, 142
Kumamoto (*Ken*), 5, 31 (Map 11), 33, 68–9 (Table 3.1)
Kumamoto (*Shi*), 68–9 (Table 3.1)
Kurashiki (*Shi*), 83, 121, 139
Kure (*Shi*) (former naval base near Hiroshima), 125
Kuril Island Chain, 4, 6, 7 (Map 3)
*Kuroshio* (Black, or Japan current), 19, 19 (Map 10)
Kurume (*Shi*), 68–9 (Table 3.1
Kushiro (*Shi*), 78
Kuwana (*Shi*), 68–9 (Table 3.1)

Kwantung Army (1931), 117
Kyoto (*Fu* and *Shi*), 31 (Map 11), 32–3, 41, 60–4, 68–9 (Table 3.1), 71, 75, 79, 84–5, 92, 102–3, 121, 129, 153, 155
*See also* Heiankyō
Kyoto Basin, 15, 59–60. 102
Kyushu, 4–5, 7–9, 13 (Table 1.2), 15, 26 (Table 1.4), 31 (Map 11), 37–8, 45, 59–60, 63, 74, 78, 80, 85, 90, 98–9, 101–2, 108–9, 114–15, 120, 125, 140–1, 143, 149, 157
Kyushu (Mt.) Node, 4, 8 (Map 3), 8

Labour, 29, 34, 40, 49, 53–4, 95, 99, 101–4, 106, 112, 123, 127, 131, 135, 152
Labour (shipyard), 135
*Laissez-faire* capitalism, 149
Landforms, 4–15, 7 (Map 2), 8 (Map 3), 10 (Map 4), 11 (Map 5), 43, 47, 60
*See also* Mountains; Surface features
Landholdings, 35, 37, 42, 49, 53–7, 146–7
*See also* Field consolidation; Fragmentation
Land ownership, 42–5, 48–52, 54, 56–7, 153, 154
Land reform (post-1945), 4, 42, 47, 48–53, 56–7, 146–7
*See also* Agricultural Land Law (1952); United States (occupation)
Landscapes, 1–4, 59, 66, 115, 120–1, 124, 128 (Fig. 7.2), 137–41, 139 (Fig. 7.3), 141–3, 150–3
agrarian, 1–4, 34–57, 119–20, 125, 145–7, 152
industrial and commercial, historical, 89, 94–5, 107, 109–10, 120, 123–4, 150–3
before 1945, 1–4, 107–24
contemporary, 1–4, 109–10, 120, 123–4
'man-made', 2–4, 59, 144
urban, 1–4, 59, 67, 109–10, 121–2, 148–50
historical urban, 1–4, 59, 61, 67–70
modern urban, 1–4, 119–20, 128 (Fig. 7.2), 137–41, 139 (Fig. 7.3), 141–3, 150–3,

181

*Index*

156—7, 158
Land use, 34, 37—48, 50—3, 55—6, 138—40, 141—3, 144—5, 153, 158
Latitudinal extent (of Japanese Archipelago), 17
Leisure, 145
Life expectancy, 136
Line of dislocation, 6, 7 (Map 2), 8
  *See also* 'Median Dislocation Line'
'Living national treasure', 95
Lockwood, W. W., 116, 118
Lowlands, 4, 10—15, 40—1, 66
  *See also* Alluvial plains; Plains

*Machi* (town), 32
  *See also* Cho
Machinery industry, 115—16, 118, 127—8, 131—4, 135, 140—3
Major Industries Control Law (1931), 117
Maebashi (*Shi*), 68—9 (Table 3.1)
Maeda (*samurai* family of Toyama), 42
Manchuria, 117, 118
Manufacturing (ancient), 67, 72, 76, 89—91. 94, 95—6, 101—4, 150—3
  boatbuilding, 108—10, 111, 116 (Fig. 6.3)
  firearms, 108—9, 111, 122
  handicrafts, 89, 94, 95—6, 99, 101—4, 150—1
  iron and metal products, 94, 108—11, 115—16, 128—31, 135—6, 157
  leather goods, 95, 150
  swords, 94, 96—7, 105, 150
Manufacturing (Modern), 86—7, 112—24, 125—43
Manufacturing specialisation
  Edo period, 71, 75, 95, 102—4, 105—6, 145—6
  modern, 108—11, 113—20, 123—4, 126—30, 132—3, 134—40, 142, 150—3
Market towns (*ichiba-machi*), 61, 74, 76, 89
Marugame (*Shi*), 68—9 (Table 3.1)
'*Masu-komi*' (mass-communications), 26
Matsue (*Shi*), 68—9 (Table 3.1), 121
Matsumoto (*Shi*), 68—9 (Table 3.1)
Matsushita Electric Corporation (Panasonic, or National), 133
  *See also* Electrical equipment; Precision industries
Matsuyama (*Shi*), 68—9 (Table 3.1), 125, 154
Matsusaka (*Shi*), 68—9 (Table 3.1)
Mazda Motors Corporation, 128 (Fig. 7.2)
'Median Dislocation Line', 6, 7 (Map 2). 8
  *See also* Line of dislocation
'Megalopolis', 30, 139
  *See also* 'Tokaido Megalopolis'
Meiji period (1868—1911), 56, 58—9, 59, 79—80, 111, 112—15, 145, 151
Meiji Restoration, 56, 58, 79, 84, 88, 106, 107—8, 108—11, 112—15
  *See also* Restoration
Merchant class (Edo period), 40—1, 99, 103—4
  *See also* Social classes
Metals and metallurgy, 17, 85, 94—7, 102, 108—11, 114—17, 125, 127—31, 139—40, 142—3
Mie (*Ken*), 31 (Map 11), 33, 68—9 (Table 3.1), 138
Mihara (*Shi*), 68—9 (Table 3.1)
Mikimoto Kokichi (founder of cultured pearl industry), 20
Military class (*samurai*), 42, 59, 72, 99, 102—3, 112
  *See also* Social classes
Minamoto and Hojo Regencies, 61
Mining and quarrying, 15, 17, 94, 111, 115, 120, 125, 127, 134, 137
  coal, 15, 17, 111, 120, 133, 134
  copper, 17, 95, 120
  cement, 19, 136
  iron, 94, 115, 122
  limestone, 120
  other, 127
Minolta Corporation, 129
  *See also* Optical goods; Photographic industry; Precision industries
*Miso* (fermented beanpaste), 104
Mito (*Shi* and *Han*), 68—9 (Table 3.1), 108, 110, 151
Mitsubishi, 115, 116 (Fig. 6.3)
  *See also* Zaibatsu; Monopolies
Mitsui, 100, 115, 120
  *See also* Zaibatsu; Monopolies
Miyagi (*Ken*), 31 (Map 11), 33, 68—9 Table 3.1)
Miyazaki (*Ken*), 31 (Map 11)
Modernisation, 3—4, 34—5, 42—53, 56—7, 79—84, 86, 89—93, 125—43
Mongol invasions (13th century), 59
Monopolies, 100—1, 104—5, 113—14,

115, 120, 123–4, 127,
    131–3, 151
  of *han* in Edo period, 100–1, 151
  modern, 100, 105, 113, 120, 123–4,
    133
  post-World War II, 127, 133
Monsoon, 10–15, 35–6, 104
  *See also* East Asian Monsoon
Morioka (*Shi*), 68–9 (Table 3.1)
Motor vehicles, 16, 81–2, 89–93,
    135, 138, 140, 142, 162
  *See also* Transportation (land)
Mountains, 4–8, 9 (Fig. 1.2), 11
    (Map 5), 13 (Table 1.2)
  *See also* Landforms; Surface features
*Mura* (village), 32
  *See also* Son
Muroran (*Shi*), 78, 115
Musashino Upland (northwestern
    Kanto), 42

Nada (town, now part of Kobe, famous
    for *sake*-brewing, 102
Nagano (*Ken*), 31 (Map 11), 40, 68–9
    (Table 3.1), 115, 129–30
Nagano (*Shi*), 74, 88
Nagasaki (*Ken* and *Shi*), 31 (Map 11),
    71–2, 87, 108–9, 115, 116
    (Fig. 6.3)
Nagoya (*Shi*), 16 (Map 8), 18 (Map 9),
    27 (Table 1.5), 68–9 (Table
    3.1), 67, 71, 74–5, 84, 87,
    92, 102, 115, 138, 140,
    154–5
Nanao (*Shi*), 109
Nara (*Ken*), 31 (Map 11), 33
Nara (*Shi*), 60–1, 75, 88, 121, 148,
    153
  *See also* Heijōkyō
Nara period (AD 710–84), 59–60,
    75, 78
Nara Basin, 15, 59–60
National Planning Board (1930s), 155
National Police Reserve (*Keisatsu
    Yobitai*), 130
National security, 119, 130
National Security Force (*Hoantai*), 130
National Self-Defence Force (*Jietai*), 130
Natural resources, 17–22, 102–3, 111,
    115, 117, 124, 125, 134–5,
    145, 150
  *See also* Human resources; raw
    materials
Nature (Japanese attitudes toward), 144
Nayoro Basin (north-central Hokkaido),
    45

New Comprehensive National Develop-
    ment Programmes (1962–
    73), 155–6
New industrial cities and special indus-
    trial zones, 19, 138–9, 155
    (Map 13), 156 (Map 14),
    156–7
New Sanyo Line (JNR), 80
New Tokaido Line (JNR), 15–16, 80,
    91–2
New Town, 89–90, 138
  *See also Danchi*
Niigata (*Ken* and *Shi*), 5, 10, 17, 31
    (Map 11), 68–9 (Table 3.1),
    87, 121–2, 140
Niigata Plain, 40–1
Niihama (*Shi*) (manufacturing city,
    northern Shikoku), 115, 120,
    125
Nishinomiya (*Shi*), 120, 139
Nissan Motors Corporation, 133
Nobeoka (*Shi*) (manufacturing city)
    (company town), north-
    eastern Kyushu), 115, 120,
    139
Nobi (Nagoya-Chukyo) Plain, 14, 41
Nodes (mountain), 4, 6, 7–8, 8, (Map 3)
North America, 12, 58
North China Plain, 12–13, 14 (Map 7)
Noto Peninsula (Map 1), 6, 9, 125, 141
Nuclear arms, 121, 137

Obama (*Shi*), 68–9 (Table 3.1)
Occupation, 47–8, 50, 122, 127–30,
    146–7, 152
  *See also* United States (occupation)
Ocean currents, 19–20. 19 (Map 10)
Oceania, 118
Oceanic monsoon, 12
  *See also* Pacific maritime air-mass
Oda Nobunaga (*Shōgun,* 16th century),
    63
Odawara (*Shi*), 68–9 (Table 3.1), 78
Ogaki (*Shi*), 68–9 (Table 3.1)
Oita (*Ken* and *Shi*), 31 (Map 11, 68–9
    (Table 3.1), 139
Okaya (*Shi*) (textile city converted to
    photo-optical manufacturing),
    129–30, 138
Okayama (*Ken*), 31 (Map 11), 33, 51,
    68–9 (Table 3.1), 83, 125,
    139, 155 (Map 13)
Okayama (*Shi*), 15, 68–9 (Table 3.1),
    71, 83, 125
Okazaki (*Shi*), 68–9 (Table 3.1)
Okinawa (*Ken*), 2, 5, 26 (Table 1.4)

# Index

'Old Japan', 2, 4, 40, 78, 79, 91, 113
Oligarchy, 108, 117, 123
　See also Meiji Oligarchs
Omuta (*Shi*) (coal-mining and manufacturing city, northern Kyushu), 115
'One castle to a province' (Edo period law), 66
Onin War (AD 1467–77), 63
*Onsen* (hot-springs), 74, 76, 88–9, 93
　See also spas
Optical goods industry, 129–30, 135, 138
　See also Japan Optical Corporation; Minolta; Photographic industry; Precision industries; Yashica; Zeiss
Orogeny (mountain-building), 5
Osaka, 27 (Table 1.5), 31 (Map 11), 32–33, 38, 41, 61–2, 62 (Fig. 3.1), 64, 65 (Fig. 3.2), 67, 68–9 (Table 3.1). 71–2, 74, 75, 77–9, 84, 87, 89–90, 92, 101, 103, 115, 120, 129, 138, 138–40, 150, 155
　castle, 64, 65 (Fig. 3.2)
　*fu*, 31 (Map 11), 32–3, 68–9 (Table 3.1), 115
　*shi*, 4, 27 (Table 1.5), 32, 61–2, 62 (Fig. 3.1), 63 (Map 12), 64, 65 (Fig. 3.2), 67. 68–9 (Table 3.1), 71–2, 74, 77–9, 84, 87, 89, 92, 101, 103, 120, 129, 138, 140, 155
Osaka Bay, 9, 41, 59, 75, 138
Otaru (*Shi*) (outport for Sapporo), 78
Ou district, 6, 71
Ou Mountain Range, 6
Outports, 61–2, 72–3, 91, 124
　See also Yokohama, Kobe
*Oyashio* (cold current), 19 (Map 10)

Pacific maritime air-mass, 12
Pacific rim, 5
Paddy (rice grain), 11, 34–46, 53, 54, 55–7, 99, 144–8
　See also Irrigated rice; Rice
Paper and pulp, 21, 75, 132, 136, 150
'Part-time farming', 29, 54, 56, 99, 127, 142
Peat bogs, 42
Perry, Matthew C., 77, 91, 98, 102, 108, 132
Petro-chemical industry, 17, 19, 132, 133–4, 135, 138–9, 142–3, 157
Petroleum, 17, 19, 80, 134–5, 137, 157
Philosophy, 119, 144–5
　See also Confucianism; Zen
Photographic industry, 129, 135, 138, 153
　See also Japan Optical Corporation; Minolta; Optical goods; Precision industries; Yashica; Zeiss
'Pioneer cities' (of Hokkaido), 78, 91
Piracy, 61, 72, 76, 94–5, 97, 105
Plains, aggradational, 4, 10–15, 40, 55, 59, 146, 163 (Footnote 3)
　See also Lowlands, or *plains by name*
Planning, 38, 60, 137–8, 143, 153–62
Plastics industry, 135–6
Politics and urbanisation, 32–3, 58–9, 61–4, 64–72, 75–6, 86–7, 91–3, 98–9, 105–6, 128, 154–8
Politics and political parties, 45, 48–50, 50, 56, 112, 128, 144, 152, 154, 157–8
Population, 22–3, 23–5, 26, 26 (Tab. 1.4), 27 (Tab. 1.5), 28 (Tab. 1.6), 29 (Tab. 1.7), 29, 30, 40, 48 (Tab. 2.1), 48, 49, 99, 105, 112–13, 118, 150
　density, 23–5, 40, 150, 159
　growth, 22–3, 24 (Tab. 1.3, 45, 48 (Tab. 2.1), 119, 148, 150
　rural, 40, 48 (Tab. 2.1), 48–9, 56, 99, 105, 117, 119
　urban, 25–6, 26 (Tab. 1.4), 27 (Tab. 1.5), 28 (Tab. 1.6), 29 (Tab. 1.7), 45, 56, 63–7, 68–9 (Tab. 3.1), 71–2, 157
'Port City', 139
　See also Kobe
Ports (*minato-machi*), 9–10, 61–3, 72, 76, 77–8, 87, 91–2, 123, 139, 157
　foreign trade
　　pre-modern, 61–2, 72, 97–8
　　modern, 72, 77–8, 87, 92, 115–23, 125, 139, 157
　strategic, 90, 93, 149
　internal, 72, 76, 87, 98
Portuguese traders, 63
Post, or stage towns (*shukuba-machi*), 72–3, 76, 87–8
Precision industries, 128–30. 142
　electrical equipment, 120, 125, 132–3, 135, 138, 153, 159; *see also* Electrical industry;

*Index*

Hitachi; Matsushita Electric Corporation; Toshiba
optical goods, 128–30, 135, 153;
  *see also* Japan Optical Corporation; Minolta; Photographic industry; Yashica; Zeiss
sewing machines, 128
Prefectural universities, 85
Prefecture (*Ken*), 32–3
Price supports, 33
Primogeniture, 43, 54, 70
Printing and publishing, 21, 85
Privacy, 145
Private railways, 80, 84, 91, 124
  *See also* Inter-urban railways; Japanese National Railways; Railways; Underground railways
Privateering *see* Piracy
Propane gas, 22
Proto-caucasoid, 59
  *See also* Ainu
Public Housing Corporation, 90
Public nuisance, 134, 140–1, 143, 150, 158, 161
  *See also* Environmental pollution; *Kōgai*
Public works, 154

Railways, 15–16, 16 (Map 8), 18 (Map 9), 79–84, 87–91, 91–2, 114
  nationalisation of railways, 80
  private railways, 80, 84, 91, 124
  railway electrification, 17, 80–1, 84, 89, 124
  railway gauge, 80
  *See also* Inter-urban railways; Japanese National Railways; Private railways; Underground railways
Range-and-Township system, 38, 60
  *See also jō-ri;* cadastral (rectilinear) survey
Raw materials, 15, 17–22, 94–5, 102–3, 115, 118, 124, 133–4, 137, 157
  raw material imports, 115, 133–4, 137, 157
  *See also* Human resources; Natural resources
Rayon and acetate staple, 136
Reclaimed lands, 41–2, 54, 55, 120, 125, 138–9, 139 (Fig. 7.3)
Reclamation, 15, 41–2, 45, 55, 125

Reconstruction (post-World War II), 122, 127–31, 141–2
Rectangular grid, 38, 60, 153
  *See also Jō-ri* system; Range-and-Township system; Cadastral (rectilinear) survey
Reischauer, E. O., 79, 99, 103, 104, 113–14, 122
Religion, 22, 58–9, 74, 88, 98
  *See also* Buddhism; Christianity; Shingo; Tenri Sect; Zen
Religious centres (*monzen-machi*), 67, 74, 76, 88, 92, 159
Rents, 43, 48–9
Reparations (post-1945), 127
Resort towns, 74, 88–9, 93
  *See also Onsen-machi;* Spas
Restoration, 56, 58, 79, 84, 88, 106, 107–8, 108–11, 112–15
  *See also* Meiji Restoration
Reverberatory furnaces (for iron production, 1850s), 108
*Ria* coast, 9
Rice, 11, 34, 35–46, 52–3, 55–7, 67, 71, 76, 150
  contemporary supply, 34, 52–3, 55–6, 147
  storage in Edo period, 71
  trade and economic importance in Edo period, 41–2, 45, 52–3, 56, 67, 76, 78, 99, 150
  *See also* Dry or upland rice; Irrigated rice; Paddy
Rice cycle, 34, 35–6, 53, 55–6
River valleys, 10, 21, 70
  *See also* individual rivers
Roads, 16, 18 (Map 9), 47, 63 (Map 12), 67–8, 78–9, 85, 92, 127, 149
  *See also Gokaidō;* Transportation (land)
Rocks, 6
Russian presence (19th century), 78, 91
Russo-Japanese War (1904–5), 114, 123
Ryukyu Archipelago, 4

Sacred treasures (mirror, jewel, sword), 96
Sado Island, 5
Saga (*Ken* and *Shi*(, 31 (Map 11), 33, 68–9 (Table 3.1), 108
Sagami Bay, 9
Saijo (*Shi*), 125
Saitama (*Ken*), 31 (Map 11), 33, 88
Sakai (*Shi*), 41, 61–3, 63 (Map 12), 65, 67, 71, 76, 97, 139

185

*Index*

See also 'Free cities'
*Sake* (rice brew, mildly alcoholic), 75, 85, 102—3, 150
Sakahalin, 6, 91
   See also Karafuto
*Samurai*, 42, 70, 72, 99, 102—3, 107, 112
   See also Social classes
*Sankinkotai* system (known also as the 'hostage system', or 'system of alternate attendance'), 67, 70, 72, 76, 79, 87—8, 98
Sano (*Shi*), 68—9 (Table 3.1)
Sansom, Sir George, 62, 146
Sanyo Trunk line (JNR), 83
   See also Railways
Sapporo (*Shi*), 12, 78, 84, 85, 113, 159
Sasebo (*Shi*) (Kyushu naval base), 90
*Sashimi* (sliced raw fish), 20
   See also Food
Satellite cities, 61, 89, 93, 120, 124
   See also Bed towns; dormitory towns
Satsuma (*han* of southern Kyushu), 108, 110
   See also Kagoshima
Savings (as a Japanese cultural attribute), 131
Sea of Japan, 5, 9, 12, 41, 125, 141
Second World War *see* World War II
Seed-beds (nursery beds for rice seedlings), 36, 53
Sendai (*Shi*), 9, 12, 41, 68—9 (Table 3.1), 71, 85, 88, 100 (Fig. 5.1), 109 (Fig. 6.1), 121 (Fig. 6.4), 126 (Fig. 7.1), 149 (Fig. 8.3), 154
Sendai Plain, 6, 12, 14—15
*Sengoku* period (late 15th to early 17th centuries), 38—40
Senri 'New Town', 89—90
Sericulture, 67, 75, 95, 102, 115
   See also Silk industry; Textiles (silk)
Service industries, 92, 123, 148
Seto (*Shi*), 75
*Setomono* (pottery, ceramics), 67, 75, 102, 136, 150
*Setonaikai* (Seto Inland Sea), 4—5, 6, 9, 12, 38, 59, 72, 78, 109, 125, 138—9, 139 (Fig. 7.3), 143, 149—50, 157
*Setouchi* (regional name for area around *Setonaikai*), 38, 125, 141
Settlement, 22—3, 30—3, 43, 48, 58—60, 61—75, 75—6, 84—93, 120
Sewage systems, 154

Sheldon, C. D., 101
*Shi* (city), 32
Shibata (*Shi*), 68—9 (Table 3.1)
Shiga (*Ken*), 31 (Map 11), 33, 68—9 (Table 3.1)
Shikoku, 4, 5 (Fig. 1.1), 6, 7 (Map 2), 13 (Map 6), 26 (Table 1.4), 31 (Map 11), 38, 125, 139, 141
Shimane (*Ken*), 31 (Map 11), 68—9 (Table 3.1)
Shimbashi (as early Tokyo rail terminus), 79
Shimonoseki (*Shi*), 87
Shimonoseki Strait, 125
*Shinden* (new paddy fields), 42
Shinto, 59, 74, 88
   See also Religion
Shipbuilding industry, 72, 85, 108—11, 116 (Fig. 6.3), 122, 125, 130, 132, 135, 142—3, 150—1, 153, 157
Shizuoka (*Ken* and *Shi*), 31 (Map 11), 33, 68—9 (Table 3.1), 71
*Shoen* period (c. 9th to 15th centuries), 38—40
*Shogun* (supreme military commander in feudal period), 63—4, 66—7, 97, 98, 100, 108, 110
   See also Shōgunate
Shōnai Plain, 41
Showa period (1926 to present), 56
*Shoyu* (soy sauce), 85
Silk industry, 67, 75, 95, 102, 115, 116—18, 123, 129, 132, 136, 138, 151
   See also Sericulture; textiles (silk)
Sino-Japanese War (1894—5), 45, 114
Slope land, 11 (Map 5), 12, 13 (Table 1.2), 15, 55
Social classes (Edo period), 42—3, 70, 79, 99, 146—8
Social control (1930s), 117—20, 123
Social reform, 56—7, 59, 107, 112—13, 119, 123—4, 152
Social welfare, 119—20, 123—4, 127, 140—1, 142—3, 147, 153—4, 157—8
Socialist party, 50
   See also Politics and Political parties
Soils, 35, 37, 53—4, 146
Sony corporation, 133
South America, 118
South Okayama, 139, 157
   See also Kurashiki (*Shi*); New industrial cities

*Index*

Southeast Asia, 75
Soviet Union, 135—6
Spain, 15
Spas (*onsen-machi*), 74, 76, 88—9, 93
    *See also* Hot springs; *Onsen*
Stage towns *see* Post towns
'Strassendorf' type settlements, 42, 72
Strategic centres, 61, 71
Strategic industry, 59, 75, 93, 94—7,
    102, 105, 108—11, 113—14,
    117-20. 122—4, 150
Styles of living, 26—9, 43, 46—8, 54,
    59, 71—2, 81, 119, 124, 134,
    142, 144—5, 146—8, 150
Sugar refining, 114
Sulphur, 17, 95
Sumitomo, 100, 115, 120
    *See also Zaibatsu;* Monopolies
*Supa* (super-markets), 54, 81
Surface features, 5—12, 36, 43, 47, 60
    *See also* Landforms; Mountains
Suruga Bay, 9
Suwa (Lake), 129
Sweden, 15, 135
'Switzerland of Asia', 130
Swords and sword-making, 75, 94—7,
    150
    *See also* Manufacturing (ancient)

Tablelands, 37, 42
Taika reforms (AD 645—701), 39, 61
Taira (former *Shi*), 68—9 (Table 3.1)
Taisho period (1912—25), 56, 108
Taiwan, 131
Takada (former *Shi*), 69—70 (Table 3.1)
Takamatsu (*Shi*), 68—9 (Table 3.1)
Takaoka (*Shi*), 68—9 (Table 3.1), 121
Takasaki (*Shi*), 68—9 (Table 3.1)
Takarazuka (*Shi*), 74
Takayama (*Shi*), 121
*Tale of Genji* (11th-century novel), 61
T'ang period (of China), 60, 75
Taoism, 144
Tariff autonomy, 111
Taxes, 43, 49, 99, 112
Technology and technological change,
    51—3, 115—16, 132—4,
    151—3
Television, 47, 85, 127, 135
Tenancy, 43, 45, 48—53, 56, 146—7
Tenri (*Shi* and religious sect), 88
Terminal department stores, 84, 92
Terraces and terracing, 37, 38
Textiles
    after World War II, 125, 128, 130—1,
        132, 136, 138, 150—1, 159

before World War II, 89, 111, 113—
        18, 120, 122—3, 132, 150—1
    cotton, 67, 75, 102, 111, 113—18,
        122—3, 125, 128, 130—1,
        132, 136, 150
    Edo period, 67, 75, 95, 102, 111,
        122, 150—1
    other, 78, 85, 95, 102, 111, 113,
        115—18, 120, 122—3, 125,
        128, 130—1, 132, 136
    silk, 75, 95, 102, 115—18, 123,
        132, 136, 138, 151
*To* (metropolitan prefecture of Tokyo),
    32
Tochigi (*Ken*), 31 (Map 11), 33, 68—9
    (Table 3.1)
Tohoku (region made up of various
    prefectures of northeastern
    Honshu), 26 (Table 1.4), 31
    (Map 11), 40—1, 45, 50, 85,
    115, 125, 127, 141, 157
Tokai district, 31 (Map 11), 70, 86,
    125, 139, 141, 143, 149
Tokaido 'megapolis', *see* Megalopolis
Tokaido (road and region), 30, 70,
    80—1, 86, 125, 139, 141,
    143, 149
Tokugawa clan, 63—7, 76, 77, 79,
    97—8
Tokugawa (Edo) period (1600—1867),
    22, 40—2, 43, 56, 63—7,
    76, 77, 89, 97—101, 101—4,
    105—6, 107—11, 145, 148,
    150—1
Tokushima (*Ken* and *Shi*), 31 (Map 11),
    68—9 (Table 3.1)
Tokyo (*To* and *Shi*), 4, 15, 16 (Map 8),
    27 (Table 1.5), 29, 30—3,
    68—9 (Table 3.1), 71—2, 77,
    79—84, 92, 109—13, 120,
    129, 138—40, 150, 153—4,
    159—60
Tokyo Bay, 9, 139
Tokyo Institute for Municipal Research,
    160
Tokyo Olympiad (1964), 159
*Tonari-gumi* (neighbourhood associa-
    tions), 129
Tosa (*han*, southern Shikoku), 108
Toshiba Electric Corporation, 133
Tottori (*Ken* and *Shi*), 31 (Map 11),
    68—9 (Table 3.1)
Tourist centres and tourism, 33, 54,
    60—1, 87, 88—9
Toyoda automatic loom, 115—16
Toyama Bay, 9

187

*Index*

Toyama (*Ken* and *Shi*), 15, 31 (Map 11), 42, 68–9 (Table 3.1)
Toyota Motors Corporation, 133, 140
Toyota (*Shi*), 140, 143
Trade
   contemporary, 126, 128–31, 132, 134, 137, 141–2, 151–3
   domestic, Edo period, 42, 67, 71, 72, 76, 94, 97–102, 105–6, 150
   modern, 85, 114–18, 122–4, 126, 128–31, 132–4, 141–2, 151–3
   pre-Tokugawa period, 61–2, 71–2, 94–7, 105, 150
Trade imbalances (1930s, other), 117, 118, 124, 126, 153
Traffic cities, 67, 90–1
*Transhumance*, 37
Transportation, 15–16, 16 (Map 8), 18 (Map 9), 63 (Map 12), 66, 67–70, 78–84, 87–8, 91–2, 97–8, 114, 116, 126–7, 148–9, 161–2
   air, 16, 81, 92
   land (rail), 15–16, 79–84, 88–90, 91–2, 114, 149, 161–2
   land (road), 16, 67–70, 78–9, 91, 97–8, 126–7
   sea, 41, 61–3, 72, 78, 81, 87–114
Tsu (*Shi*), 68–9 (Table 3.1)
Tsukushi Plain (northern Kyushu), 15, 38
Tsugaru (old shipbuilding centre), 109
Tsugaru Strait, 15, 42
Tsuruga (*Shi*) (site of nuclear 'accident', 1981), 137
Tsurumi (Kanto area satellite town), 120
Tsuruoka (*Shi*), 68–9 (Table 3.1)
Tsuyama (*Shi*), 68–9 (Table 3.1)
Typhoons, 12, 41

Ueda (*Shi*), 68–9 (Table 3.1)
Ueno *or* Iga Ueno (*Shi*), 68–9 (Table 3.1)
*Unagi domburi* flavoured, broiled eel on a bed of hot rice), 20
   *See also* Food
Underground railways, 84, 89, 92
United Nations, 131, 160
United States, 77, 90–1, 107, 118–19, 121, 124, 127–8, 130–1, 137, 142, 146–7, 149, 151–2, 161–2
   influences in agriculture, 45, 47, 48, 145

occupation (1945–52), 47–8, 122, 127–8, 130–1, 146–7, 152
trade relationships, 116, 118–19, 124, 126, 128–31, 149
Uplands, 37, 42
Uprising (peasant), 42, 56, 147
Urawa (*Shi*), 88
Urban
   function, 61–3, 64–75, 75–6, 77, 84–93, 97, 115
   growth, 58, 60–75, 75–6, 77–93, 97–9, 104–5, 108–9, 115, 120, 124, 138, 148–50, 153–8
   hierarchy, 15, 58, 61–4, 64–76, 84–93, 148–50
Urban agriculture, 29, 54, 56, 147–8, 153
Urban-industrial regions, 17, 115, 120, 138–40, 149–50, 153–8
USSR *see* Soviet Union
Utsunomiya (*Shi*), 68–9 (Table 3.1), 88
Uwajima (*Shi*), 5 (Fig. 1.1), 68–9 (Table 3.1)

Values (cultural), 43, 58, 95, 96, 101, 119, 123, 130–1, 136, 144–5, 147–8, 152
Villages (*son*), 30 (Table 1.8), 32, 43
Viticulture, 51–2, 136
Volcanoes and vulcanism, 4, 4–8, 42

*Waju* (polders at head of Ise Bay), 41
Wakamatsu (Aizu) (*Shi*), 68–9 (Table 3.1)
Wakasa Bay, 9
Wakayama (*Ken* and *Shi*), 31 (Map 11) 33, 68–9 (Table 3.1), 139
Wakkanai (*Shi*) (airbase of northern Hokkaido), 90
Water resources and supply, 19, 38, 51, 102, 145, 154
Weapons of war, 59, 75, 95–7, 102, 105, 108, 110, 113–14, 125, 142, 150–2
Weights and measures, 97
Wells and springs, 38
Wet rice *see* Paddy; irrigated rice
'White Heron castle' (Himeji-*Shi*), 86
Wine and Whisky making, 51–2, 136
World War I, 45, 89, 114, 123, 131, 151
World War II, 47–8, 50, 56, 85, 90, 93, 116–17, 119, 121–2, 124, 125, 127, 129, 131,

# Index

137, 151–2, 154
Writing, 21, 59

Yamagata (*Ken* and *Shi*), 31 (Map 11), 41, 68–9 (Table 3.1), 127
Yamaguchi (*Ken* and *Shi*), 31 (Map 11) 33, 68–9 (Table 3.1), 78, 90, 125
Yamanashi (*Ken*), 31 (Map 11), 38, 40, 51–2, 68–9 (Table 3.1)
Yanagawa (*Shi*), 68–9 (Table 3.1)
Yangtse lowlands (China), 37
Yashica Company, 129–30
*See also* Optical goods; Photographic industry; Precision industries; Zeiss
Yatsushiro Bay (Kyushu), 41
Yawata (former *Shi*), 114
Yawata Iron Works (1901), 114–15
Yen devaluation, 117, 124

Yodo (*gawa*-River), 62, 62 (Fig. 3.1)
Yokkaichi (*Shi*), 74, 138
Yokohama (*Shi*), 77, 79, 83, 91, 109, 120, 139
Yokosuka (*Shi*) (naval base), 90, 109
Yonezawa (*Shi*), 68–9 (Table 3.1)
Yoshikawa Eiji (modern historical novelist), 61
Yoshimune (*shōgun* of Tokugawa Clan, c. 1721), 100

*Zaibatsu*, 100, 105, 113, 115, 120, 123–4, 127, 133, 152
*See also* Mitsubishi; Mitsui; Sumitomo; Monopolies
*Zaibatsu*-dissolution, 133, 152
Zeiss (German Optical firm), 130
*Zen*, 96–7, 144
Zoning, 153

189